PHARMACEUTICALS AND NUTRACEUTICALS FROM FISH AND FISH WASTES

PHARMACEUTICALS AND NUTRACEUTICALS FROM FISH AND FISH WASTES

Ramasamy Santhanam, PhD
Santhanam Ramesh, PhD
Subramanian Nivedhitha, PhD
Subbiah Balasundari, PhD

AAP | APPLE ACADEMIC PRESS

First edition published 2022

Apple Academic Press Inc.
1265 Goldenrod Circle, NE,
Palm Bay, FL 32905 USA
4164 Lakeshore Road, Burlington,
ON, L7L 1A4 Canada

CRC Press
6000 Broken Sound Parkway NW,
Suite 300, Boca Raton, FL 33487-2742 USA
4 Park Square, Milton Park, Abingdon,
Oxon OX14 4RN

Library and Archives Canada Cataloguing in Publication

Title: Pharmaceuticals and nutraceuticals from fish and fish wastes / Ramasamy Santhanam, PhD, Santhanam Ramesh, PhD, Subramanian Nivedhitha, PhD, Subbiah Balasundari, PhD.
Names: Santhanam, Ramasamy, 1946- author. | Ramesh, Santhanam, 1979- author. | Nivedhitha, Subramanian, author. | Balasundari, Subbiah, author.
Description: First edition. | Includes bibliographical references and index.
Identifiers: Canadiana (print) 20210299215 | Canadiana (ebook) 2021029924X | ISBN 9781774630105 (hardcover) | ISBN 9781774638767 (softcover) | ISBN 9781003180548 (ebook)
Subjects: LCSH: Fisheries—By-products. | LCSH: Fishes—Utilization. | LCSH: Drug development. | LCSH: Functional foods.
Classification: LCC TP996.F5 S26 2023 | DDC 338.3/727—dc23

Library of Congress Cataloging-in-Publication Data

Names: Santhanam, Ramasamy, 1946- author. | Ramesh, Santhanam, 1979- author. | Nivedhitha, Subramanian, author. | Balasundari, Subbiah, author.
Title: Pharmaceuticals and nutraceuticals from fish and fish wastes / Ramasamy Santhanam, PhD, Santhanam Ramesh, PhD, Subramanian Nivedhitha, PhD, Subbiah Balasundari, PhD.
Description: First edition. | Palm Bay, FL : Apple Academic Press, 2023. | Includes bibliographical references and index. | Summary: "This comprehensive book brings together experts from both the marine science and pharmacy disciplines to relay important aspects on the pharmaceutical and nutraceutical values of 175 species of bony and cartilaginous fishes as well as the uses of fish processing byproducts and wastes. Presented in an easy-to-read style, the volume provides precise identification of freshwater and marine fishes possessing pharmaceutical and nutraceutical compounds along with over 180 photographs. Aspects covered include biology, ecology, diagnostic features, and pharmaceutical and nutraceutical compounds along with their activities for each of the fish included. The book details the bioactive compounds, including fish muscle proteins, peptides, collagen and gelatin, fish oil, etc., from such species, as well as the bioactive peptides that are derived from various fish muscle proteins, which have various biological activities, including cardio protective, antihypertensive, anticancer, anti-diabetic, antibacterial, anticoagulant, anti-inflammatory, and antioxidant activities. Also discussed are the nutritional benefits of fish consumption, which are largely due to the presence of proteins, unsaturated essential fatty acids, minerals, and vitamins. The waste products obtained during fish processing are also a potential source of bioactive peptides that can be used as a source of nitrogen and amino acids, which have immunomodulatory, antibacterial, antitrombotic, and antihypertensive properties. This volume provides the information needed to tap into these vast pharmaceutical and nutraceutical benefits. Pharmaceuticals and Nutraceuticals from Fish and Fish Wastes will be of great use for students and researchers of disciplines such as pharmaceutical sciences, marine/fisheries sciences, marine microbiology, and marine biotechnology. It will also be a standard reference for libraries of colleges and universities and a guide for pharmaceutical companies involved in the development of new drugs from fishes and their wastes"-- Provided by publisher.
Identifiers: LCCN 2021040738 (print) | LCCN 2021040739 (ebook) | ISBN 9781774630105 (hardcover) | ISBN 9781774638767 (paperback) | ISBN 9781003180548 (ebk)
Subjects: LCSH: Fisheries--By-products. | Drug development. | Functional foods.
Classification: LCC TP996.F5 S26 2023 (print) | LCC TP996.F5 (ebook) | DDC 338.3/727--dc23/eng/20211108
LC record available at https://lccn.loc.gov/2021040738
LC ebook record available at https://lccn.loc.gov/2021040739

ISBN: 978-1-77463-010-5 (hbk)
ISBN: 978-1-77463-876-7 (pbk)
ISBN: 978-1-00318-054-8 (ebk)

About the Authors

Ramasamy Santhanam, PhD, is the former Dean of the Fisheries College and Research Institute, Tamil Nadu Dr. J. Jayalalithaa Fisheries University, India. He has 50 years of teaching and research experience in marine sciences, and his fields of specialization include marine biology and fisheries environment. He is currently serving as a fisheries expert for various governmental and non-governmental organizations in India and abroad. Dr. Santhanam has so far published 30 books on various aspects of marine life, marine plankton, and aquaculture. He was a member of the American Fisheries Society, United States; World Aquaculture Society, United States; Global Fisheries Ecosystem Management Network (GFEMN), United States; and the IUCN's Commission on Ecosystem Management, Switzerland.

Santhanam Ramesh, PhD, is Professor in the Department of Pharmaceutics at the School of Pharmacy at Sri Balaji Vidyapeeth (Deemed to be University), Pondicherry, India. He obtained his PhD from Jawaharlal Nehru Technological University, Hyderabad, India. He has 15 years of teaching and research experience in pharmaceutical sciences, and his fields of specialization include natural products, pharmaceutical nanotechnology, biomaterials, and marine wastes of pharmaceutical importance. Dr. Ramesh is also a visiting professor in the Department of Pharmacology, North-Caucasian State Humanitarian and Technological Academy (North-Caucasian State University), Cherkessk, Russia. To his credit, he has 10 books published with internationally reputed publishers. He is a member of the International Society for Pharmacoeconomics and Outcome Research, (ISPOR), USA; member of the British Society for Nanomedicine, UK; and associate member of the Academic Pharmacy Group of Royal Pharmaceutical Society, London.

Subramanian Nivedhitha, PhD, is Associate Professor and Head, Department of Pharmacognosy, Ratnam Institute of Pharmacy, Nellore, AP, India. She obtained her PhD from Jawaharlal Nehru Technological University, Hyderabad, India. Her fields of specialization include pharmacognosy, phytochemistry, and pharmacology. She has 13 years of teaching and research experience and to her credit. She has six research papers published in peer-reviewed journals.

Subbiah Balasundari, PhD, is the Dean, Dr. M. G. R. Fisheries College and Research Institute, Tamil Nadu Dr. J. Jayalalithaa Fisheries University, Thalainayeru, India. She has 25 years of teaching and research experience in fish processing. She has developed a number of contemporary value-added fish products and disseminated the technologies to fish processing industries of India. She has completed six projects through state and national level funding agencies and established various demonstration units in fisheries enterprise for the benefit of stakeholders. She is a member of AFSIB, Agricultural Scientific Tamil Society, SOFT, and WAS.

Contents

Abbreviations

5-LOX	5-lipoxygenase
ACE	angiotensin-converting enzyme
ASC	acid-soluble collagen
CH	crude homogenate
DH	degree of hydrolysis
DHA	docosahexaenoic acid
DPPH	1,1-diphenyl-2-picrylhydrazyl
DPP-IV	dipeptidyl-peptidase IV
dw	dry weight
ECM	epigonal conditioned medium
EPA	eicosapentaenoic acid
Epi-1	epinecidin-1
FPHs	fish protein hydrolysates
GF	gray fat
HFFD	high-fat and fructose diet
HGFs	human gingival fibroblasts
IL	interleukin
MBC	minimum bactericidal concentration
MEC	minimal effective concentration
MIC	minimum inhibitory concentration
MRSA	methicillin-resistant *Staphylococcus aureus*
MUFA	mono-unsaturated fatty acids
PSC	pepsin-soluble collagen
PUA	polyunsaturated aldehyde
PUFA	polyunsaturated fatty acid
SFAs	saturated fatty acids
SOD	superoxide dismutase
STZ	streptozotocin
TC	total cholesterol
TG	triglyceride
TNF	tumor necrosis factor
UGP	undigested goby protein
VRE	vancomycin-resistant enterococci

Foreword

I am very much delighted to write the foreword for this book, *Pharmaceuticals and Nutraceuticals from Fish and Fish Wastes.* Fish, nature's superfood, plays an important role in human nutrition. Low fat, lean fish meat is considered better than mutton, beef, and even poultry for human health. Consuming 100 g of small pelagic fish such as sardines or anchovies twice a week will more than cover the needs of omega-3s for a person in coastal areas. In many developing countries, fish is the only source of animal protein and micronutrients. As of 2019, Asia ranked first in the per capita fish consumption (25.1 kg), followed by North America (23.7 kg), Europe (21.6 kg), South America (10.7 kg), and Africa (9.8 kg).

The fish processing industry has become a major exporter of seafood and marine products. During processing, a significant amount of waste (20–80%) is generated, and this waste has been estimated at about 25% of the fish catch. To overcome the environmental impact of these wastes, there is an urgent need to convert these wastes into a potential source of pharmaceuticals and nutraceuticals.

This comprehensive book written by experts from both the fisheries science and pharmacy disciplines deals with the pharmaceutical and nutraceutical values of fishes as well as fish processing by-products and wastes. The volume provides the information needed to tap into these vast pharmaceutical and nutraceutical benefits.

This book will be of great use for students and researchers of disciplines such as marine/fisheries sciences, pharmaceutical sciences, and marine biotechnology. It will also become a standard reference for libraries of colleges and universities and as a guide for pharmaceutical companies involved in the development of new drugs from fish and their wastes.

I have no doubt that the constituents of fisheries sector will derive great benefit from this book. I congratulate the lead author Dr. Ramasamy Santhanam and his team for their sincere efforts in bringing out this timely publication.

—Dr. G. Sugumar
Vice Chancellor
Tamil Nadu Dr. J. Jayalalithaa Fisheries University
Nagapattinam 611002, India

Preface

The nutritional aspects of fish consumption are largely due to the presence of proteins, unsaturated essential fatty acids, minerals (e.g., calcium, iron, selenium, and zinc), and vitamins, namely vitamin A, B3, B6, B12, E, and D. Fish are also the reservoirs of structurally diverse bioactive materials such as protein and peptides, polyunsaturated fatty acids (PUFA), polyphenols, and so on with numerous health benefits.

Natural medicinal products have been used for millennia for the treatment of multiple ailments in several countries like China and India. Although many of such products have been superseded by modern pharmaceutical approaches, currently there is a resurgence in interest in the use of natural products particularly from aquatic life. In this context, fish and their bioactive compounds such as muscle proteins, peptides, collagen and gelatin, fish oil, and so on are known to possess potential pharmaceutical and nutraceutical values. Among these bioactive compounds, peptides derived from various fish muscle proteins assume greater significance owing to their various bioactivities including cardioprotective, antihypertensive, anticancer, antidiabetic, antibacterial, anticoagulant, anti-inflammatory, and antioxidant activities. Apart from the consumed parts of fish, the waste products obtained during fish processing such as skin, scales, fins, and viscera including liver; and muscle of underutilized fish are also potential sources of oils, minerals, enzymes, and bioactive peptides. Among these peptides, nitrogen and amino acids have a lot of potential functions such as immunomodulatory, antibacterial, antithrombotic, and antihypertensive activities. Hence, the bioactive compounds of fish and their processing wastes may be of great use to biomedical and food industries.

Though some books are presently available on marine natural products, a comprehensive book on pharmaceutically and nutraceutically important fishes and their wastes is hitherto wanting. Keeping this in consideration, this publication is being brought out for the first time by bringing together the experts of both marine science and pharmacy disciplines. Aspects relating to pharmaceutical and nutraceutical values of 175 species of bony and cartilaginous fishes and fish processing by-products are dealt with.

It is hoped that the present publication, written in an easy-to-read style, will be of great use for students and researchers of disciplines such as pharmaceutical sciences, marine/fisheries sciences, marine microbiology, and marine biotechnology, and as a standard reference for all the libraries of colleges and universities, and as a guide for pharmaceutical companies involved in the development of new drugs from fishes and their wastes.

We are highly indebted to Dr. G. Sugumar, Vice-Chancellor, Tamil Nadu Dr. J. Jayalalithaa Fisheries University, Nagapattinam, India, for his kind foreword. Thanks are also due to Dr. K. Venkataramanujam, former Dean, Fisheries College and Research Institute of Tamil Nadu Dr. J. Jayalalithaa Fisheries University, India, for his valued comments and suggestions on the manuscript. I sincerely thank all my international friends who were very kind enough to permit their images for the present purpose. The secretarial assistance rendered by Mrs. Albin Panimalar Ramesh is also gratefully acknowledged.

—**The Authors**

CHAPTER 1

Introduction

ABSTRACT

This chapter deals with the importance of fish in human nutrition; marine pharmaceutical and nutraceutical compounds and their present status; and on the pharmaceutical and nutraceutical values of fish wastes and their by-products, such as fish protein powder, fish protein hydrolyzates, and fish oils.

Among aquatic biotopes, seas and oceans rank first owing to their vast area compared to freshwater habitats. While the seas and oceans cover 70 % of the earth's surface, the freshwater biotopes viz. rivers, lakes, ponds, and other related habitats cover just 1% of the Earth's surface. Further, the seas and oceans occupy 90% of the biosphere. Among the vertebrates, fish comprise 28,600 species. Of these species, 95% are bony fishes, mostly teleosts; about 50 species of agnathans (jawless fishes); and 800 are cartilaginous fish species. Marine biotopes harbor 58% of fishes while 41% are freshwater fish, and the remaining 1.0% is diadromous (Ullah and T. Ahmad, 2014).

1.1 FISH AND HUMAN NUTRITION

Fish, the nature's superfood, plays an important role in human nutrition. Fish is not only a source of proteins and healthy fats but also a unique source of essential nutrients, including long-chain omega-3 fatty acids, iodine, vitamin D, and minerals such as calcium, magnesium, iron, sodium, iodine, and phosphorus. Protein is present in fish diets in the form of simple proteins with different essential amino acids. They have a fine taste, are easily digestible, and have a high growth, which upholds value. The fish fats grant more energy as compared to other animals. Vitamin D, which is generally missing from cattle meat, is present in the flesh of fish. Of the required vitamins, 14% can be attained from fish use. Different studies have shown that low fat, lean fish meat is better than mutton, beef, and even poultry for human health. Small fish species are the main source of protein and most of the fat-soluble vitamins. Specifically in many developing countries of the world, the rural people are completely relying on fish food sources solely (Ullah and Ahmad, 2014). Fish plays an important role in fighting hunger and malnutrition. The multiple benefits of fatty fish high in omega-3s and small fish eaten whole containing nutrients in the skin and bones clearly illustrate seafood's irreplaceable nutritional value. In many developing countries, fish is the main or only source of animal protein and is essential for providing micronutrients to vulnerable populations. For instance, goiter is found in areas where iodized salt is unavailable, but the consumption of fish and the natural iodine it contains could help reduce these cases. Eating fish once or twice a week may also reduce the risk of stroke, depression, Alzheimer's disease, and other chronic conditions. Consuming one hundred grams of small pelagic fish, such as sardines or anchovies, once a week will more than cover the needs of omega-3 for a person in coastal areas (Abbey, et al., 2017). Further, for inland people, two meals a week of most carps will be adequate, and no fish oil is needed in their feed in order to become a good source of beneficial omega-3 oils (Toppe, 2014).

1.1.1 FISH CONSUMPTION IN THE WORLD

Fish formed 16.6% of the animal protein in the human diet (Ghaly et al., 2013). As on 2019, Asia ranked first in the per capita fish consumption (25.1 kg) followed by North America (23.7 kg), Europe (21.6 kg), South America (10./kg), and Africa (9.8 kg) (Shahbandeh, 2019).

Marine pharmaceutical and nutraceutical compounds and their present status: Among the marine species, invertebrates like sponges and cnidarians have been extensively explored in the last decades for their novel bioactive natural products or pharmaceutical compounds. On the other hand, fish are the little-studied group and their pharmaceutical and nutraceutical properties are largely limited to their by-products such as collagens/gelatin, oil, bones, powder, and so on, derived from their wastes. Pharmaceuticals work only after the onset of diseases and nutraceuticals help in the prevention and retardation of diseases. The key difference between nutraceutical and pharmaceutical products is that nutraceutical products are not required to undergo the same testing and regulations as pharmaceutical products. Pharmaceuticals are provided under prescription from certified physicians for purchase while nutraceuticals do not require the surpervision of health professionals.

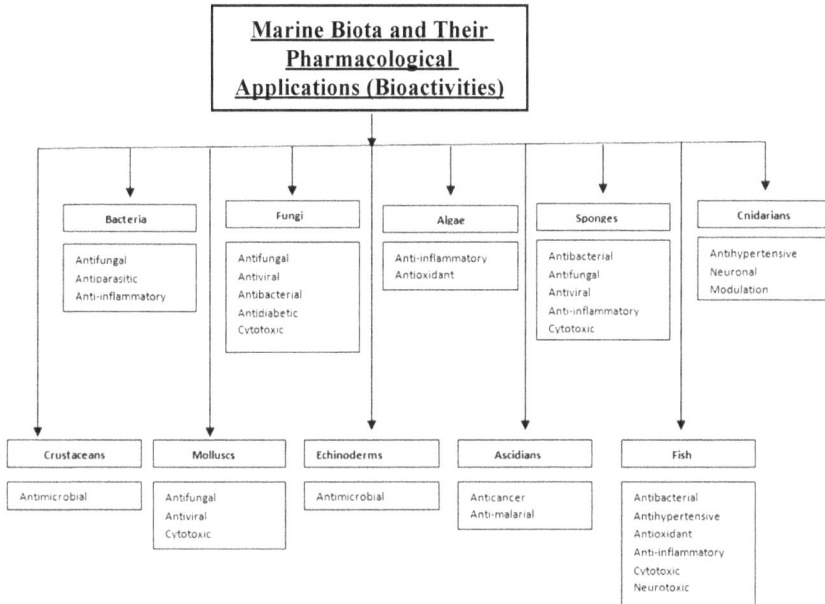

Marine Biota and Their Pharmacological Applications (Bioactivities)

Bacteria	Fungi	Algae	Sponges	Cnidarians
Antifungal Antiparasitic Anti-inflammatory	Antifungal Antiviral Antibacterial Antidiabetic Cytotoxic	Anti-inflammatory Antioxidant	Antibacterial Antifungal Antiviral Anti-inflammatory Cytotoxic	Antihypertensive Neuronal Modulation

Crustaceans	Molluscs	Echinoderms	Ascidians	Fish
Antimicrobial	Antifungal Antiviral Cytotoxic	Antimicrobial	Anticancer Anti-malarial	Antibacterial Antihypertensive Antioxidant Anti-inflammatory Cytotoxic Neurotoxic

1.2 FISH WASTES AND BY-PRODUCTS

"Fish wastes" include fish species having no or low commercial value; undersized commercial species; and fins, heads, skin, and viscera discarded by fish-processing industries. The problem of unused/underutilized fish wastes which are estimated at about 25% of the fish catch has increased in the last years and it has become a global concern. To overcome the environmental impact that fish wastes, there is an urgent need to convert these wastes into potential source of pharmaceuticals and nutraceuticals. From the unused/underutilized fish wastes, a significant amount of proteins (enzymes, collagen), protein hydrolysates, lipids, and so on, could be produced with wide biotechnological applications (Caruso, 2016).

1.2.1   FISH PROTEIN POWDER

It is a potential source of several minerals such as potassium, phosphorus, and magnesium, and essential amino acids.

1.2.2   FISH PROTEIN HYDROLYZATES

Although fish protein hydrolyzate is used as a protein supplement, fish collagen hydrolysate produced from the skins of wild deep-sea ocean fish such as cod, haddock, and pollock serves as a therapeutic agent in the treatment of osteoarthritis and osteoporosis. Further, the peptides derived from the fish protein hydrolyzates have been reported to possess antioxidant, antihypertensive, immunomodulatory, and antimicrobial peptides.

1.2.3   FISH OILS

The world fish oil production has been reported to range from 5.5 to 7.5 million tons/year. The notable fish oil-producing countries are Peru and Chile, which contribute to the 52% and 13% (https://www.fitday.com/fitness-articles/nutrition/vitamins-minerals/8-advantages-of-fish-fatty-acids.html). The health benefits of fish oils include the following:

a)   Anticancer property: Consumption of fish oil containing omega-3 fatty acids reduces the risk of cancer.
b)   Cardiovascular effects: Eating fatty fishes twice a week improves cardiovascular health by reducing the risk of stroke and heart disease.
c)   Reduces dementia risk: Weekly consumption of sea fish lowers the risk of developing both dementia and Alzheimer's disease in elderly people.
d)   Antidiabetic property: Eating fatty fish helps blood sugar levels in people with diabetes.
e)   Anti-inflammatory activity: Regular eating of fatty fish helps in relieving inflammatory conditions like autoimmune diseases, rheumatoid arthritis, and on.
f)   Brain and eye health: Omega-3 fatty acids of fish oils have contributed to the health of both brain tissue and the retina in humans.

g) Anti-asthma: Fatty fish possess anti-asthma properties especially in children.

1.3   FUTURE CONSIDERATIONS

Advanced fish preservation and processing methods are the need of the hour to mitigate the problem of fish wastes. Further, biotechnological applications have to be popularized to convert the fish wastes into value-added by-products.

KEYWORDS

- **human nutrition**
- **pharmaceutical values**
- **nutraceutical values**
- **fish wastes**
- **by-products**
- **fish protein powder**
- **fish oils**

CHAPTER 2

Pharmaceuticals and Nutraceuticals from Fish and Their Activities

ABSTRACT

This chapter deals with the biomedically important bony fishes and elas-mobranchs along with their profile. Although the profile has the common name, global distribution, and biology and ecology of individual species, the biomedical aspects deal with the pharmaceutical and nutraceutical compounds derived from each fish species along with their bioactivities.

2.1 BONY FISHES (PHYLUM: CHORDATA; SUBPHYLUM: VERTEBRATA; CLASS: OSTEICTHYES)

2.1.1 STURGEONS (ORDER: ACIPENSERIFORMES; FAMILY: ACIPENSERIDAE)

Acipenser persicus Borodin, 1897

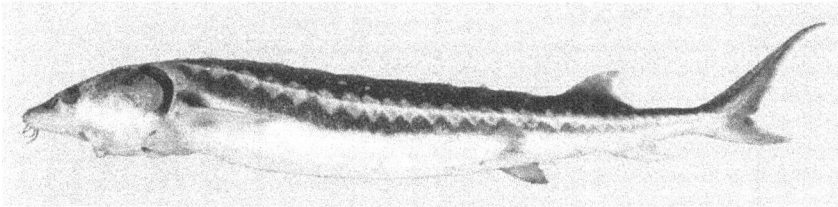

Source: A. Abdoli. https://commons.wikimedia.org/w/index.php?curid=29530383

Common name(s): Persian sturgeon

Global distribution: Temperate; Eurasia: Caspian basin; eastern Black Sea

Habitat: This demersal, anadromous species largely inhabits coastal and estuarine (brackish water) zones at the sea.

Maximum length and weight: 255 cm; 70.0 kg

Food and feeding: This species feeds mainly on gammarid amphipods, mysid shrimp, and chironomid flies.

Uses: Commercial fisheries exist for this species. The cooked fillets of this species serve as a potential source of proteins for 10–12 years old and adults (Alipour et al., 2010). Further, the values of concentration (microgram per gram, wet weight) of heavy metals (such as Cr, 0.27; Pb, <0.01; As, <0.01; and Co, <0.01) of the caviar—roe of this species were found to be significantly lower than adverse level for human consumption when compared with FAO/WHO permissible limits (Hosseini et al., 2013).

Pharmaceutical and nutraceutical compounds and activities

Antihypertensive and anticoagulant properties: The polysaccharide extracted from the cartilage of this species showed the highest ACE (angiotensin-converting enzyme) inhibitory activity (85.7%) at the highest concentration of 1 mg/mL. Sturgeon sulfated polysaccharide induced significant anticoagulant activities, as indicated by the activated partial thromboplastin time (APTT) and thrombin time (TT) indexes, at concentrations between 10 and 100 µg/mL. The clotting times were also significantly prolonged—about 2.07 and 1.3 folds greater than control in terms of APTT and TT, respectively (Karimzadeh et al., 2018).

Acipenser schrenckii **Brandt, 1869**

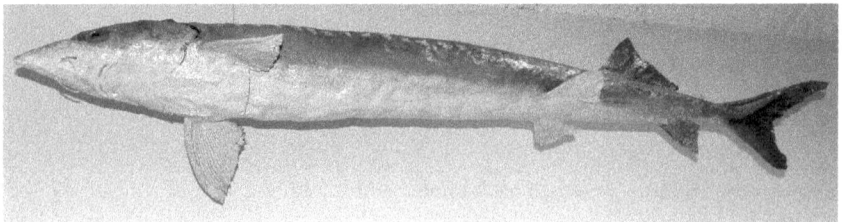

Source: A. C. Tatarinov, https://commons.wikimedia.org/wiki/File:Acipenser_schrenckii.jpg

Common name(s): Amur sturgeon

Global distribution: Two morphs, namely, brown and gray are endemic to Amur River system of Asia; Sea of Japan

Habitat: Adults are demersal and anadromous; marine and brackish water areas; prefers sandy or stony bottom

Maximum length and weight: 300 cm; 190 kg

Food and feeding: Feeds on benthic organisms

Uses: It is a highly prized fish; commercial fisheries exist for this species.

Pharmaceutical and nutraceutical compounds and activities

Antioxidant and cryoprotective effects: The skin gelatin hydrolysates of this species were found to be effective in preventing lipid oxidation as evidenced by the lower thiobarbituric acid-reactive substances formation. These gelatin hydrolysates were able to retard protein oxidation as indicated by the retarded protein carbonyl formation and lower loss in sulfhydryl content. Further, the oligopeptides in gelatin hydrolysates more likely contributed to the cryoprotective effect (Nikoo et al., 2015).

2.1.2 FRESHWATER EELS (ORDER: ANGUILLIFORMES; FAMILY: ANGUILLIDAE)

Anguilla anguilla **Temminck & Schlegel, 1846**

Source: Supino, Felice (1916) Pesci d'Acqua Dolce d'Italia, Milan: Ulrico Hoepli, Editore Libraio della Real Casa. Courtesy of Freshwater and Marine Image Bank.

Common name(s): European eel

Global distribution: Subtropical; Japan, Korea, China, Taiwan, and Vietnam; and northern the Philippines

Habitat: It inhabits marine, freshwater, and brackish water areas; depth range 1–400 m; demersal and catadromous

Maximum length and weight: 122 cm; 6.6 kg

Food and feeding: Adults of this species feed mainly on crustaceans and fish.

Uses: It is the most expensive food fish in Japan. It is also used in Chinese medicine.

Pharmaceutical and nutraceutical compounds and activities

Antibacterial and hemolytic activities: The mucus of this fish showed antibacterial activity against *Vibrio fluvialis*, *Vibrio parahaemolyticus*, *Vibrio alginolyticus*, and *Staphylococcus aureus* with a mean diameter of inhibition 0.80 cm. The blood serum of this species showed similar activity against only *Vibrio alginolyticus* with 0.95 cm. Further, the blood serum of this species also showed hemolytic properties (Caruso et al., 2014).

Anguilla japonica Temminck & Schlegel, 1846

Source: James Carson Brevoort, 1818-1887. Public domain.
https://commons.wikimedia.org/w/index.php?curid=21633032

Common name(s): Japanese eel

Global distribution: Subtropical; Asia: Japan, East China Sea, Taiwan, Korea, China, and the Philippines

Habitat: It occurs in marine, freshwater, and brackish water areas; depth range 1–400 m; demersal and catadromous

Maximum length and weight: 100.8 cm; 1.9 kg

Food and feeding: This species feeds mainly on crustaceans and fish.

Uses: Commercial fisheries and aquaculture exist for this species.

Pharmaceutical and nutraceutical compounds and activities

Antimicrobial activity: Its peptide, Cathelicidin 1, showed antimicrobial activity (Donati et al., 2011; https://www.uniprot.org/uniprot/J7GIU8).
Antioxidant activity: The diethyl ether extracts of the skin and flesh of this species showed significant 1,1-diphenyl-2-picrylhydrazyl (DPPH) radical scavenging activity with 89.2% and 61.5%, respectively (Ekanayake et al., 2005).

2.1.3   CONGER EELS (ORDER: ANGUILLIFORMES; FAMILY: CONGRIDAE)

Conger myriaster Brevoort, 1856

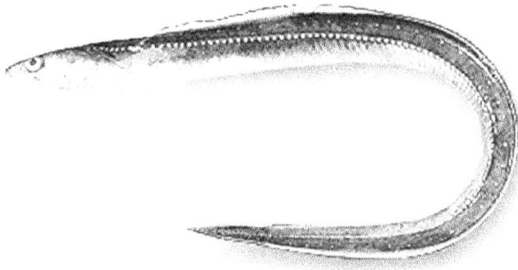

Source: http://www.fao.org/fishery/species/2996/en

Common name(s): Whitespotted conger; conger eel

Global distribution: Temperate; northwest Pacific: East China Sea, Korean Peninsula, and Japan

Habitat: It is a bathydemersal and oceanodromous species inhabiting shallow bottom sand and mud at a depth range of 320–830 m

Maximum length and weight: 152.0 cm; 12.6 kg

Food and feeding: Feeds on shrimps, fishes, and cephalopods

Uses: It is the most relished fish among the congrids. It is captured and cultured for fishery in Japan.

Pharmaceutical and nutraceutical compounds and activities

Antioxidant activity: The diethyl ether extracts of the skin and flesh of this species showed significant DPPH radical scavenging activity with 23.0% and 17.1%, respectively (Ekanayake et al., 2005). Ranathunga et al. (2005) reported that its muscle peptides showed scavenging of hydroxyl radicals and carbon-centered radicals with IC_{50} values of 74.1 µM and 78.5 µM, respectively.

2.1.4   MORAY EELS (ORDER: ANGUILLIFORMES; FAMILY: MURAENIDAE)

***Strophidon sathete* (Hamilton, 1822)** (*=Evenchelys macrurus*)

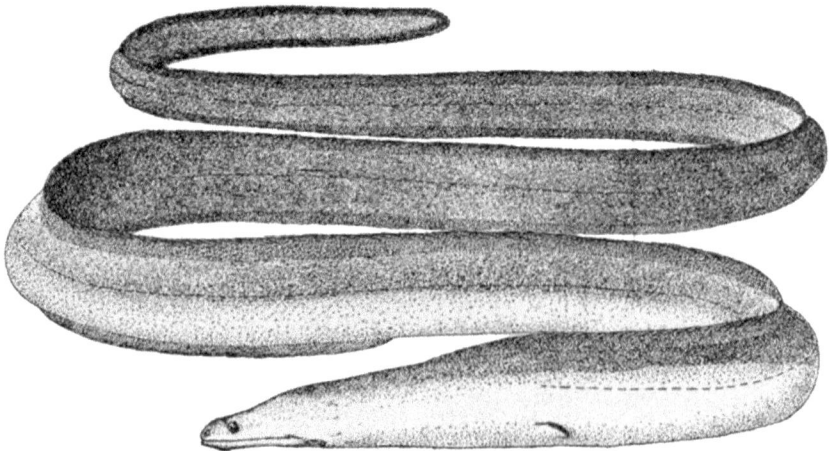

Source: Bleeker - The fishes of the Indo-Australian Archipelago p.293. Public domain. https://commons.wikimedia.org/w/index.php?curid=40262310

Common name(s): Slender giant moray, Gangetic moray

Global distribution: Red Sea and East Africa to the western Pacific

Habitat: It is an amphidromous fish inhabiting both muddy ocean bottoms and estuarine areas at a depth of 15 m.

Maximum length: 400 cm

Food and feeding: Feeds mainly on a variety of small fishes and crustaceans.

Uses: It has minor commercial fisheries.

Pharmaceutical and nutraceutical compounds and activities

The type I collagen derived from the outer skin waste of this species could be a source in the development of new drugs due to its closeness in denaturing temperature to mammalian collagen. Further, the gel and film-forming capability of this collagen-containing implant standard antibiotic has been proved to be a suitable drug delivering system (Veeruraj et al., 2012).

2.1.5 LIZARDFISHES (ORDER: AULOPIFORMES; FAMILY: SYNODONTIDAE)

Saurida elongata (Temminck & Schlegel, 1846)

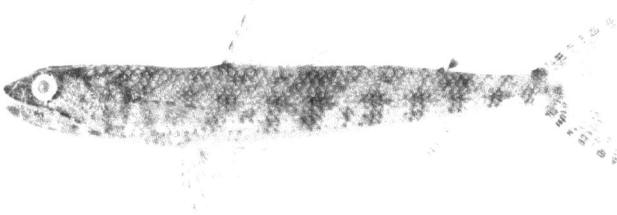

Source: Bernard Dupont / Flickr. License: CC BY Attribution-ShareAlike. https://fishesofaustralia.net.au/home/species/2786

Common name(s): Slender lizardfish, elongate lizardfish

Global distribution: Temperate; northwest Pacific: Japan to the northern South China Sea

Habitat: This marine, demersal inhabits sandy bottoms in shallow water at a depth range of 20–100 m.

Maximum length: 50.0 cm

Food and feeding: Lizardfish is generally carnivorous and predacious feeding mainly on fishes followed by crustaceans and mollusks.

Uses: Minor commercial fisheries exist for this species.

Pharmaceutical and nutraceutical compounds and activities

Anti-anemia activity: The enzymatic hydrolysate of this species showed significant anti-anemia activity on tested anemia models induced by blood loss (Dong et al., 2005).

ACE-inhibitory activity: The peptide with the amino acid sequence Arg–Val–Cys–Leu–Pro derived from the muscle protein hydrolysate of this species showed ACE-inhibitory activity with an IC_{50} value of 175 µM. This study suggests that peptide from this lizardfish hydrolysate may be beneficial as antihypertensive compound for use as functional food ingredients or pharmaceutical item against hypertension (Wu et al., 2015).

Sun et al. (2017) reported that the purified peptide with the amino acid sequence, Arg–Tyr–Arg–Pro derived from the protein hydrolysate of this species showed significant ACE-inhibitory activity with an IC_{50} value of 52 µM.

2.1.6   FLYING FISH (ORDER: BELONIFORMES; FAMILY: EXOCOETIDAE)

Exocoetus volitans Linnaeus, 1758

Source: http://www.fishbase.org Public domain.

Common name(s): Tropical two-wing flyingfish, blue flyingfish, common flyingfish

Global distribution: Subtropical; western Pacific: Japan, Marshall Islands, the Philippines, Australia, and Tahiti; eastern Pacific: Mexico to central Chile; Galapagos and Hawaii

Habitat: This species is pelagic and is inhabiting both near and far from the coast. It forms schools.

Maximum length: 30.0 cm

Food and feeding: It is feeding mainly on crustaceans and planktonic animals.

Uses: Commercial fisheries exist for this species.

Pharmaceutical and nutraceutical compounds and activities

Antioxidant and antiproliferative activity: The peptic protein hydrolysate from the backbone of this species showed significant free radical scavenging and lipid peroxidation inhibition activity. Further, the purified extract exerted a significant antiproliferative activity against Hep G2 cell lines (Naqash and Nazeer, 2011).

2.1.7 HERRINGS AND SARDINES (ORDER: CLUPEIFORMES; FAMILY: CLUPEIDAE)

Clupea pallasii Valenciennes, 1847

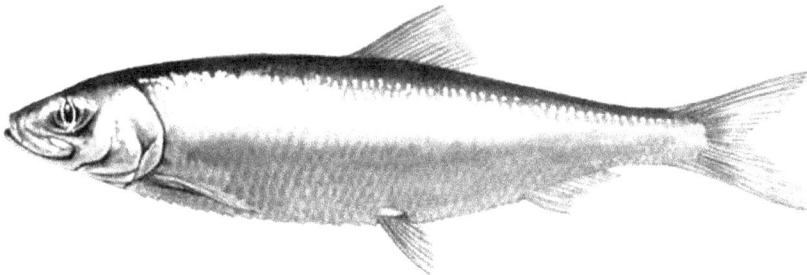

Common name(s): Pacific herring

Global distribution: Temperate; Arctic: White Sea; western Pacific: Anadyr Bay, Kamchatka, Aleutian Islands, Japan, and Korea; eastern Pacific: Kent Peninsula and Beaufort Sea southward to northern Baja California, Mexico

Habitat: This pelagic-neritic species lives in marine, freshwater, and brackish water areas. This schooling species is nonmigratory and has a depth range of 0–475 m.

Maximum length and weight: 46.0 cm

Food and feeding: Adults of this species prey on crustaceans and small fishes and young mainly on decapod and mollusk larvae.

Uses: This species has high commercial fisheries and it is also a game fish with recreational value.

Pharmaceutical and nutraceutical compounds and activities

Antioxidative activity: The purified two peptides (identified as Leu–His–Asp–Glu–Leu–Thr (MW = 726.35 Da) and Lys–Glu–Glu–Lys–Phe–Glu (MW = 808.40 Da), derived from the protein hydrolysates of this species showed hydroxyl radical scavenging activity and DPPH radical scavenging activity and the IC_{50} values of these activities are given in the following table.

Antioxidant activities of *Clupea pallasii* (IC_{50} (mg/mL)).

Sample	DPPH radical scavenging activity	Hydroxyl radical scavenging activity
1	5.1	4.6
2	4.4	3.8

Source: Wang et al. (2019).

Clupea harengus Linnaeus, 1758

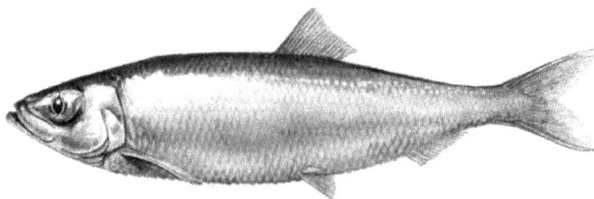

Common name(s): Atlantic herring

Global distribution: It is a very common species of North Atlantic.

Habitat: It is a benthopelagic, oceanodromous species inhabiting marine and brackish water areas; forms large schools; depth range 0–364 m.

Maximum length and weight: 45 cm; 1.1 kg

Food and feeding: It is feeding mainly on crustaceans such as copepods and krill and small fish.

Uses: This game fish has highly commercial fisheries.

Pharmaceutical and nutraceutical compounds and activities

Bioactivities: The peptides derived from the skin of this species showed the following bioactivities.

Bioactivity of skin peptides of *C. harengus*.

Bioactivity	%
Antioxidant	39.9
Cardiovascular system	51.3
Immunomodulatory	7.8
Opioid antagonist	0.05

Source: Pampanina et al. (2012).

Antioxidative activity: The protein hydrolysate derived from this species showed antioxidative activity with a percentage value of 44.4% (Sathivel et al., 2013).

Nutraceutical properties: This species has been reported to possess significant amount (11%) of the nutritionally eicosapentaenoic acid (EPA) (Jensen et al., 2007).

Sardina pilchardus **(Walbaum, 1792)**

Common name(s): Moroccan sardine; European pilchard

Global distribution: Subtropical; Northeast Atlantic: Mediterranean and Black Sea

Habitat: This pelagic-neritic and oceanodromous species lives in marine, freshwater, and brackish water areas and its depth range is 10–100 m. It is a schooling species.

Maximum length and weight: 27.5 cm;

Food and feeding: It mainly feeds on zooplankton such as copepods and their larvae, and phytoplankton.

Uses: Highly commercial fisheries exist for this species. This fish is also sold in salted and smoked or dried condition. Further, it is used in the production of fishmeal and as a fishing bait.

Pharmaceutical and nutraceutical compounds and activities

Antioxidant enzymes activities: The protein hydrolysates of this species exerted a potent cholesterol-lowering effect accompanied by efficient decrease in lipid peroxidation in serum and target tissues. Due to these factors, increased antioxidant enzyme activity was noticed in high-cholesterol-fed Wistar rats (Athmani et al., 2015).

ACE inhibition activity: The protein hydrolysates of this species have shown ACE inhibition activity and helped in controlling blood pressure (Galvez and Berge, 2013).

Others: The collagen peptide of this species has been reported to repair damaged cartilage when it is orally administrated (Khora, 2013).

Sardinella aurita **Valenciennes, 1847**

Common name(s): Round sardinella

Global distribution: Subtropical; Atlantic Ocean: West African coast from Gibraltar southward to Saldanha Bay in South Africa; Mediterranean Sea and Black Sea

Habitat: It is a coastal and brackish water species inhabiting depths up to 350 m, and it is schooling and strongly migratory.

Maximum length: 30 cm

Food and feeding: Adults feed mainly zooplankton, especially copepods and juveniles on phytoplankton.

Uses: Commercial fisheries exist for this species in the West African coast, Mediterranean Sea, and coasts of Venezuela and Brazil.

Pharmaceutical and nutraceutical compounds and activities

Antioxidant activities: The protein hydrolysates of this species exerted a potent cholesterol-lowering effect accompanied by efficient decrease in lipid peroxidation in serum and target tissues. Due to these factors, increased antioxidant enzyme activity was noticed in high-cholesterol-fed Wistar rats (Athmani et al., 2015).

Souissi et al. (2007) reported that the protein hydrolysates derived from the heads and viscera of this species displayed antioxidant activity with more than 50% inhibition of linoleic acid peroxidation.

Nurdiani et al. (2017) also reported on the DPP radical scavenging activity of the peptides derived from the head or viscera of this species.

The protein hydrolysates of this species derived through crude enzyme preparations from *Bacillus pumilus* A1 (SPHA1) and *Bacillus mojavensis* A21 (SPHA21); and crude enzyme extract from the viscera of this species antioxidant activity. While the SPHA21 had the highest DPPH radical scavenging activity (89% at 6 mg/mL), SPHEE exhibited the highest metal chelating activity (89% at 1 mg/mL) (Khaled et al., 2014).

Antibacterial, antioxidant, and ACE-inhibitory activities: The peptides derived from the protein hydrolysate of this species through treatment with *Bacillus subtilis* A26 proteases showed antibacterial, antioxidant, and ACE-inhibitory activities. While its peptide F2 displayed the highest antibacterial and ACE-inhibitory activities, peptide F4 showed the highest antibacterial and antioxidant activities (Jemil et al., 2017).

Sardinella longiceps Valenciennes, 1847

Common name(s): Indian oil sardine

Global distribution: Tropical; northern and western Indian Ocean; Gulf of Aden, Gulf of Oman, and Andaman Islands

Habitat: This neritic, oceanodromous species has a depth range of 20–200 m. It forms schools in coastal waters and is strongly migratory.

Maximum length and weight: 23 cm; 200 g

Food and feeding: It feeds mainly on diatoms and small crustaceans

Uses: This species has highly commercial fisheries. It is marketed fresh, dried and dried-salted; smoked and canned. It is also made into fishmeal and fish balls.

Pharmaceutical and nutraceutical compounds and activities

Antioxidant activity: The scavenging effect increased with the increasing concentration of hydrolysates was observed, these results indicate a dose-dependent activity. The protein hydrolysate of this species derived through trypsin showed significant scavenging activity, that is, 70.1% at 5 mg/mL (Jeevitha et al., 2014).

Gastroprotective activities (nutraceutical properties): The oil of this species is rich in EPA and docosahexaenoic acid (DHA) at 16% and 14%, respectively. This oil-treated Wistar rats showed significant reduction in ulcer. The oil of this species may therefore be a suitable natural source for the prevention and treatment of gastric lesions (Vishnu et al., 2017).

2.1.8 ANCHOVIES (ORDER: CLUPEIFORMES; FAMILY: ENGRAULIDAE)

Engraulis anchoita Hubbs & Marini, 1935

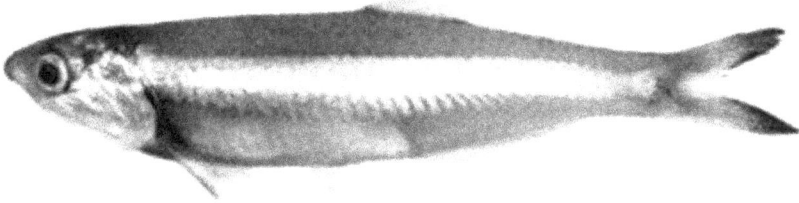

Common name(s): Argentine anchovy

Global distribution: Subtropical; Southwest Atlantic: north of Rio de Janeiro, Brazil to San Jorge Gulf, and Argentina

Habitat: This marine, pelagic-neritic species is oceanodromous and it occurs in coastal waters to about 800 km or more from the shore, forming dense schools at about 30–90 m depth in summer, but down to 100–200 m during winter.

Maximum length and weight: 17.0 cm; 25.00 g

Food and feeding: Feeds as juveniles on zooplankton, but with phytoplankton becoming increasingly important

Uses: Highly commercial fisheries exist for this species.

Pharmaceutical and nutraceutical compounds and activities

Antimicrobial properties: The protein films from this fish incorporated with 0%, 0.50%, 0.75%, and 1.50% sorbic or benzoic acids exhibited antimicrobial properties against *Escherichia coli* O157:H7, Listeria monocytogenes, Salmonella *enteritidis*, and *Staphylococcus aureus* (da Rocha et al., 2014).

Enzyme production: The crude extract of the viscera and heads from this species showed proteolytic and trypsin activity to the tune of 0.0218 U/mg. The proteolytic enzymes of this species may therefore have biotechnological applications (Lamas et al., 2017).

Engraulis encrariscolus (Linnaeus, 1758)

Common name(s): European anchovy

Global distribution: Subtropical; Eastern North and Central Atlantic; Mediterranean and Black and Azov seas

Habitat: This pelagic-neritic and oceanodromous species is found in both marine and brackish water areas; depth range 0–400 m; however, it is mainly a coastal species, forming large schools.

Maximum length and weight: 20.0 cm; 26.46 g

Food and feeding: Feeds on planktonic organisms

Uses: This species has high commercial fisheries. It is usually canned, salted, or processed, but also marketed fresh or frozen in African countries.

Pharmaceutical and nutraceutical compounds and activities

Antioxidant activity: The hexane, butanol-soluble, aqueous-soluble fraction, and precipitated crystals fractions obtained from the viscera of this species showed antioxidant activity and the values obtained by DPPH method were 964.8, 1354.1, 1522.1, and 820.3 (mmol Trolox/kg, respectively (Burgos-Hernández et al., 2016).

Antibacterial activity: The visceral extracts of this species showed antibacterial activity against pathogenic bacterial strains and the minimum inhibitory concentration (MIC) 50 values recorded with hexane and butanol extracts are given in the following table.

Antibacterial activity (MIC, mg/100 mL) of viscera fractions on pathogenic bacterial species.

	Hexane extract		Butanol extract	
	MIC50	MIC50	MIC50	MIC50
Salmonella enterica subs. *enterica* CCM 3807	6.4	6.8	6.4	6.8
Bacillus subtilis sub. *spizizenii* CCM 1999	2.1	2.4	25.5	27.1
Enterococcus faecalis CCM 4224	0.8	0.9	6.5	11.2
Shigella sonei CCM 1373	10.7	17.5	14.5	22.2
Escherichia coli CCM 3988	12.8	13.6	6.4	6.8
Staphylococcus aureus subs. *aureus* CCM 2461	3.2	3.4	25.5	27.1

Source: Burgos-Hernández et al. (2016).

Antifungal activity: The hexane and aqueous-soluble fractions of the viscera of this species showed antifungal activity against *Alternaria alternata*. The hexane fraction showed 26% inhibition of fungus growth (Burgos-Hernández et al., 2016).

Antimutagenic activity: The strong antimutagenic compounds present in the viscera of this species are believed to be responsible for its antimutagenic activity (Burgos-Hernández et al., 2016).

Engraulis japonicus Temminck & Schlegel, 1846

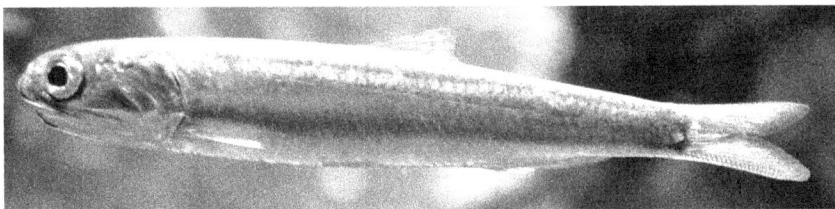

Source: Image by Kingfisher. https://creativecommons.org/licenses/by-sa/4.0/

Common name(s): Japanese anchovy

Global distribution: Temperate; western Pacific: Democratic People's Republic of Korea; Republic of Korea; China; Hong Kong; Taiwan; Japan; the Philippines; Indonesia; Russia

Habitat: This pelagic-neritic species has a depth range of 0–400 m. This oceanodromous species forms large schools. Juveniles are often found associated with the drifting seaweed.

Maximum length and weight: 18.0 cm; 45.0 g

Food and feeding: It feeds on diatoms and copepods; molluscan larvae, and fish eggs and larvae.

Uses: This species possesses commercial aquaculture and fisheries value. It is marketed fresh and salted and is often processed into fishmeal and oil.

Pharmaceutical and nutraceutical compounds and activities

Immunostimulating effect: The protein fraction obtained from the salted cooking wastes of these fish larvae showed both growth-promoting and immunostimulating effect toward cell line (Kong et al., 2002).

Calcium extraction: The salted product of this species serves as a calcium supplement (Kim et al., 2013).

Setipinna taty (Valenciennes, 1848)

Common name(s): Half-fin anchovy, scaly hairfin anchovy

Global distribution: Tropical western Indo-Pacific region

Habitat: It is found in the coastal waters and estuaries. It is also an oceano-dromous, schooling species having a depth range of 0–50 m.

Maximum length: 15.3 cm

Food and feeding: It is a zooplankton feeder. While adult fish feeds mainly on crustaceans such as Mysidacea and *Acetes chinensis*, its young feed on copepods.

Uses: Only minor commercial fisheries exist for this species.

Pharmaceutical and nutraceutical compounds and activities

Anticancer effects: The peptide derived from this species showed the amino acid sequence Tyr–Ala–Leu–Arg–Ala–His (YALRAH) and it was found to possess anticancer effects and was pro-apoptotic on PC-3 cells with an IC_{50} value of 11.1 µM (Pangestuti and Kim, 2017).

Antioxidant and antiproliferative activity: The hydrolysate of this species derived with pepsin (Hp) showed strong DPPH radical scavenging activity (ED_{50} value, 4.46 µg/mL) at a concentration of 6.80 µg/mL. Further, both the pepsin hydrolysate (Hp) and the heated pepsin hydrolysate (Hp-H) of this species displayed antiproliferative activity by inhibiting the proliferation of DU-145 human prostate cancer cell line, 1299 human lung cancer cell line and 109 human esophagus cancer cell line. The inhibitory rate of Hp-H on the DU-145 cell line was significantly higher than that of Hp at the same concentration, with an IC_{50} value of 13.7 mg/mL, about three-fold stronger than that of Hp (IC_{50}: 41.7 mg/mL). Furthermore, at a concentration of 40 mg/mL, the inhibitory rate of Hp-H (95.68%) was significantly higher than that of Hp (46.06%) with 1299 human lung cancer cells; and the IC_{50} values of Hp-H and Hp on 1299 human lung cancer cells were 25.17 and 40.28 mg/mL, respectively. The antiproliferative activity between Hp and Hp-H against different cancer cell lines is compared in the following table (Song et al., 2011).

Comparison of Antiproliferative Activity (in IC_{50} Values) between Hp and Hp-H.

Cell lines (mg/mL)	Samples	Concentration			
		5	10	20	40
U-145 human prostate cancer cell	Hp	0.2	8.4	23.4	44.4
	Hp-H	98.8	17.3	50.1	71.3
1299 Human lung cancer cell	Hp	–	4.5	20.6	46.1
	Hp-H	4.9	9.3	21.3	95.7
109 Human esophagus cancer cell	Hp	–	–	–	29.9
	Hp-H	–	–	–	56.0

Source: Song et al. (2011).

Antibacterial activity: The peptide fraction of the pepsin hydrolysate of this species exhibited antibacterial activity against *Escherichia coli* (Song et al., 2012).

Wang et al. (2019a) also reported on the antibacterial activity of synthesized peptides (HGM-Hp1, HGM-Hp2, and HGM-Hp3) derived from this species. The values of percentage inhibition of these peptides against *Escherichia coli* were reported as 42.10%, 17.93%, and 32.14%, respectively; and the value of MIC was however same (2 mg/mL) for all these peptides.

2.1.9   BARBS, CARPS, AND ROACH (ORDER: CYPRINIFORMES; FAMILY: CYPRINIDAE)

Barbodes everetti (Boulenger, 1894)

Source: Haplochromis. https://creativecommons.org/licenses/by-sa/3.0/deed.en

Common name(s): Clown barb

Global distribution: Tropical; Asia: Borneo and Sumatra

Habitat: It is freshwater, benthopelagic species found in forest streams of the foothills' zones, usually in clear, slow to fast flowing waters, preferring quieter areas along the shores; commonly found in shallow waters (less than 5–15 cm) such as puddles in the forest.

Maximum length: 15.0 cm

Food and feeding: The food items of this species include plant matter, worms, crustaceans, and insects

Uses: This species is commercially important with edible and aquarium values.

Pharmaceutical and nutraceutical compounds and activities

Antibacterial activity: The acidic and aqueous extracts of the mucus of this fish showed antibacterial activity against six human pathogenic bacterial strains. While the acidic extract showed against all the bacterial strains, the aqueous extract showed activity only on *Pseudomonas aeruginosa* ATCC 27853. The values of zone of inhibition recorded against all the five bacterial species by the acidic extract are given below.

Antibacterial activity of *Barbodes everetti* mucus extract.

Bacteria	Zone of inhibition (mm dia.)
Salmonella typhimurium	8.7
Vibrio cholerae	10.1
Staphylococcus aureus ATCC 25933	8.7
Pseudomonas aeruginosa ATCC 27853	11.1
Salmonella braenderup ATCC BAA 664	9.5

Source: Lim et al. (2018).

Catla catla (Hamilton, 1822)

Common name(s): Catla

Global distribution: Subtropical; Asian countries such as India, Bangladesh, Pakistan, Nepal, and Myanmar

Habitat: This benthopelagic and potamodromous species is found in freshwater areas such as rivers, lakes and culture ponds, and brackish water areas at a depth range of 0–5 m.

Maximum length and weight: 182 cm; 38.6 kg

Food and feeding: It is a surface and mid-water feeder. It is an omnivore, feeding mainly on phytoplankton, insects, and detritus.

Uses: It is a game fish and highly commercial fisheries and aquaculture exist for this species.

Pharmaceutical and nutraceutical compounds and activities

Antioxidant properties: The protein hydrolysates derived from this species using different proteases, namely, alcalase (A), bromelain (B), flavorzyme (F), and protamex (P) showed antioxidant properties. The values of linoleic acid peroxidation inhibition of the A, B, F, and P hydrolysates were found to be 54%, 50%, 49%, and 36%, respectively (Elavarasan et al., 2014).

Ctenopharyngodon idella (Valenciennes, 1844)

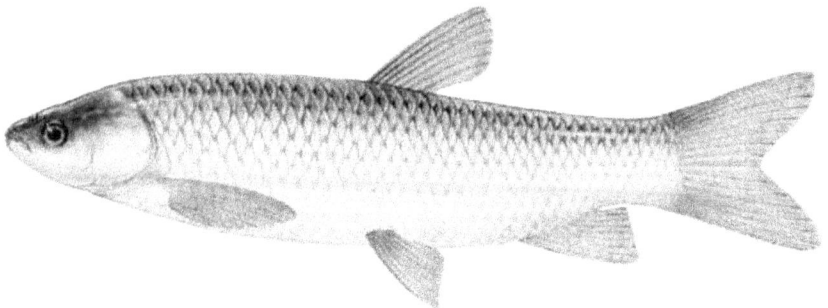

Common name(s): Grass carp

Global distribution: Asia: China to eastern Siberia

Habitat: This demersal, potamodromous species dwells in both freshwater areas such as lakes, reservoirs, ponds, and pools; and brackish water areas at a depth range of 0–30 m. Adults prefer slow-flowing or standing water bodies with vegetation.

Maximum length and weight: 150 cm; 45 kg

Food and feeding: This species feeds mainly on aquatic plants and submerged grasses.

Uses: It is a very popular aquaculture species with food value. It is also used for weed control in aquaculture systems.

Pharmaceutical and nutraceutical compounds and activities

Antioxidant activity: The three purified peptides isolated from the skin of this species showed high DPPH radical scavenging activity with IC_{50} values of 2.459, 3.634, and 6.063 mM, respectively; hydroxyl radical scavenging activity with IC_{50} values of 3.563, 2.606, and 4.241 mM, respectively; and ABTS radical scavenging activity with IC_{50} values of 0.281, 0.530, and 0.960 mM, respectively (Cai et al., 2015).

ACE-inhibitory and antihypertensive activity: The peptides extracted from the hydrolysates of this species showed ACE-inhibitory activity with an IC_{50} value of 0.692 mg/mL. Further, these peptides also displayed significant antihypertensive activity at a concentration of 100 mg/kg (Chen et al., 2016).

Cyprinus carpio Linnaeus, 1758

Common name(s): Common carp

Global distribution: Subtropical; Europe to Asia: Black, Caspian, and Aral Sea basin

Habitat: This benthopelagic, potamodromous species dwells in freshwater areas such as large turbid rivers and large, vegetated lakes, and brackish water areas.

Maximum length and weight: 120 cm; 40.1 kg

Food and feeding: This species feeds on benthic organisms and decayed plant materials.

Uses: This species has high commercial fisheries. Further, it is also a popular aquaculture and aquarium species.

Pharmaceutical and nutraceutical compounds and activities

Antioxidant effects: The defatted roe protein hydrolysates (CDRH) of this species showed DPPH scavenging activity in a dose-dependent manner. The concentrations of CDRH-30, -60 and -90 required to inhibit 50% of the free radical content were found to be 8, 14, and 12.5 mg/mL, respectively (Ghelichi et al., 2018).

Korczek et al. (2018) reported that the protein extracts (before hydrolysis) showed higher DPPH scavenging activity (129 µM/g) than their hydrolysates which showed a value of 30 µM/g. Further, the values of the ABTS scavenging ability increased in both protein extracts and hydrolysates were 30.9 and 232.3 µM/g, respectively.

ACE-inhibitory activity: The muscle hydrolysate of this species prepared after 2-h gastric digestion showed ACE-inhibitory activity with an IC_{50} value of 1.90 mg/mL (Borawska et al., 2015).

Antibacterial effects: The defatted roe protein hydrolysates of this species showed activity against Gram-positive bacterial strains, namely, *Staphylococcus aureus* (ATCC 25923), *Micrococcus luteus* (ATCC 4698), *Bacillus cereus* (ATCC 11778); and Gram-negative strain, *Escherichia coli* (ATCC 25922) (Ghelichi et al., 2018).

Immunomodulatory effects: The protein hydrolysates derived from the roe of this have been reported to significantly enhance the proliferation of spleen lymphocytes. Its pepsin hydrolysate significantly increased the splenic natural killer cell cytotoxicity, and level of serum immunoglobulin A. Further, its alcalase hydrolysate induced increase in the percentages of $CD4^+$ and $CD8^+$ cells in spleen (Chalamaiah et al., 2015a).

Anti-inflammatory effects: Its omega-3 fatty acids have been reported to lower the inflammatory bowel syndromes. Further, its good cholesterol lowers the formation of rheumatoid arthritis and osteoarthritis.

Nutritional values of common carp Health benefits of common carp *(https://www.healthbenefitstimes. com/carp-fish/).*

Hormones: Its iodine helps to balance the thyroid gland functions and hormones in the body.

Immunity enhancement: Its rich zinc content enhances immunity.

Heart functioning: Its high amount of omega-3 fatty acids prevents the formation of plaque and atherosclerosis.

Gastrointestinal health: Addition of this fish to the diet has been reported to help reduce constipation, bloating, stomach upset, and hemorrhoids in humans.

Respiratory ailments: High amount of nutrients and minerals of this carp helps to maintain the respiratory health by reducing chronic respiratory distress, bronchitis, and other illness are associated with respiratory tracts and lungs.

Bone and teeth health: Its high phosphorus content prevents the chances of osteoporosis, damaged, or weakened enamel in teeth.

Vision: It is reported that this carp food promotes the vision and lowers oxidative stress and vision deficiency.

Aging process: Antioxidants of this species have been reported to slow down the process of aging.

Sleep: Its magnesium content stimulates the release of neurotransmitters, which helps to ease nervous system for restful sleep. The intake of this carp food also helps to treat insomnia.

Cognition enhancement: Its minerals, namely, selenium and zinc, anti-oxidants, and omega-3 fatty acids stimulate neural pathways and prevent oxidative stress in the blood vessels and capillaries of brain. Further, these factors have been reported to prevent Alzheimer's disease and dementia.

Gibelion catla (Hamilton, 1822) (=*Catla catla*)

Common name(s): Indian major carp, catla

Global distribution: Africa; Asia: India, Bangladesh, Pakistan, Nepal, and Myanmar

Habitat: This benthopelagic, potamodromous species is found in freshwater areas such as rivers, lakes and culture ponds, and brackish water areas.

Maximum length and weight: 182 cm; 38.6 kg

Food and feeding: It is mainly omnivorous with juveniles feeding on aquatic and terrestrial insects, detritus, and phytoplankton.

Uses: Highly commercial fisheries exist for this species. It is also an important aquaculture species.

Pharmaceutical and nutraceutical compounds and activities

Antibacterial activity: Its crude bile displayed activity against two Gram-positive bacterial strains, namely, *Bordetella pertussis* and *Staphylococcus aureus*; and two Gram-negative bacteria, namely, *Escherichia coli* and *Vibrio cholerae* with zone of inhibition values of 13, 9, 13, and 16 mm, respectively (De et al., 2012).

Antidiabetic activity: The crude bile (100%) of this species when administered orally showed antidiabetic activity in streptozotocin (STZ)-induced diabetic mice and the percentage diabetic reduction was 74.6 (De et al., 2012).

Others: Rajani and Alka (2015) reported that the consumption of this species may cure asthma, heart diseases, inflammatory diseases, and mineral deficiency.

Hypophthalmichthys molitrix (Valenciennes, 1844)

Common name(s): Silver carp

Global distribution: Subtropical; Asia: China and eastern Siberia

Habitat: This benthopelagic, potamodromous species lives in both fresh-water and brackish water areas at a depth range of 0–20 m. It prefers warm backwaters, lakes, and flooded areas with slow current.

Maximum length and weight: 105 cm; 50.0 kg

Food and feeding: Adults are shallow water (0.5–1.0 m) feeders on phytoplankton, while larvae and small juveniles prey largely on zooplankton.

Uses: This species has commercial fisheries and aquaculture value.

Pharmaceutical and nutraceutical compounds and activities

Antioxidative properties: Experimental mice fed with the collagen hydrolysate of this species showed antioxidative enzyme activities in both serum and skin. It is suggested that this hydrolysate may serve as a dietary supplement to combat photoaging (Song et al., 2017).

Labeo rohita (Hamilton, 1822)

Common name(s): Rohu

Global distribution: Tropical; Asia: India, Bangladesh, Pakistan, Myanmar, and Nepal

Habitat: This benthopelagic species lives in both freshwater rivers and brackish water areas at a depth of 0–5 m. It is a solitary, potamodromous species burrowing occasionally.

Maximum length and weight: 200 cm; 45 kg

Food and feeding: Feeds largely on plants

Uses: It has highly commercial fisheries with aquaculture and game fish values.

Pharmaceutical and nutraceutical compounds and activities

Antibacterial activity: The crude bile of this species showed activity against Gram-positive bacterial strains such as *Bordetella pertussis* and *Staphylococcus aureus*; and Gram-negative bacteria like *Escherichia coli* and *Vibrio cholerae* with zone of inhibition values of 14, 12, 12, and 9 mm, respectively (De et al., 2012).

Antidiabetic activity: The crude bile (100%) of this species when administered orally showed antidiabetic activity in STZ-induced diabetic mice and the percentage diabetic reduction was 76 (De et al., 2012).

Antioxidant activity: Peptides of the roe protein hydrolysates of this species showed DPPH radical scavenging and ferric-reducing power activities (Galla et al., 2012).

Rutilus frisii (Kamensky, 1901)

Common name(s): Common roach, kutum

Global distribution: Temperate; Iranian Waters of the Caspian Sea

Habitat: This semi-anadromous species is found in brackish estuaries and their large, plume waters, coastal lakes, and lowland stretches of large rivers. Its depth range is 0–50 m.

Maximum length and weight: 70.0 cm; 8 kg

Food and feeding: Adults feed on benthic invertebrates such as mollusks and crabs. On the other hand, larvae and juveniles feed on zooplankton, algae, and insect larvae.

Uses: It has commercial fisheries and aquaculture values.

Pharmaceutical and nutraceutical compounds and activities

Antibacterial activity: The skin mucus extracts of both males and females this species showed activity against *Streptococcus iniae*, *Yersinia ruckeri*, *Staphylococcus aureus*, *Listeria monocytogenes*, *Pseudomonas aeruginosa*, and *Escherichia coli* (Adel et al., 2018). The values of inhibition zone diameter and MIC recorded for the different bacterial strains are given in the following table:

Bactericidal activities of skin mucus of *R. frisii.*

Bacterial strains	Zone of inhibition (mm dia.)		MIC (µg/mL)
	Male fish	Female fish	
Escherichia coli	24.3	24.0	125
Staphylococcus aureus	16.7	18.4	500
Listeria monocytogenes	20.2	22.6	250
Streptococcus iniae	16.2	17.4	>500
Pseudomonas aeruginosa	23.2	25.4	250
Yersinia ruckeri	24.7	25.1	125

Source: Adel et al. (2018).

Antifungal activity: The skin mucus extracts of this species also showed activity against *Candida albicans*, *Fusarium solani*, *Saprolegnia* sp., and *Aspergillus flavus* (Adel et al., 2018). The values of inhibition zone diameter, MIC, and minimum bactericidal concentration (MBC) recorded for the different fungal species are shown in the following table:

Antifungal activities of skin mucus of *R. frisii.*

Fungal pathogens	Zone of inhibition (mm dia.)		MIC (µg/mL)
	Male fish	Female fish	
Aspergillus flavus	16.1	16.5	>250
Candida albicans	14.2	14.5	500
Saprolegnia sp.	17.0	18.9	>125
Fusarium solani	19.2	18.6	125

Source: Adel et al. (2018).

Enzyme activities: The mucus of female fish of this species showed significant lysozyme activity than its male counterpart (Adel et al., 2018). The values of enzyme activities of both male and female fish are given in the following table:

Enzyme activities in skin mucus of *R. frisii.*

Enzyme (IU/mg)	Male fish	Female fish
Alkaline phosphatase	73.6	73.9
Esterase	2.9	3.2
Lysozyme	22.7	25.3
Protease	30.9	31.1

IU: International unit.

Source: Adel et al. (2018).

2.1.10   CODS (ORDER: GADIFORMES; FAMILY: GADIDAE)

***Gadus chalcogrammus* Pallas, 1814** (=*Theragra chalcogramma*)

Source: NOAA FishWatch. Public domain.

Common name(s): Alaska Pollack

Global distribution: Polar; North Pacific with abundance in the eastern Bering Sea

Habitat: This bentho-pelagic fish are found in both marine brackish water areas at a depth range 30–1280 m and are moving to deeper water to form schools during the day.

Maximum length and weight: 91.0 cm; 3.9 kg

Food and feeding: It mainly feeds on krill but it may also eat fishes and copepods such as adult *Acartia* and *Pseudocalanus*

Uses: Commercial fisheries exist for this species in Russian, Japanese, and Korean coasts. In Japan, this fish is sold as frozen and as fillets, fish sticks, surimi, and roe. For fishmeal and industrial products also, this fish is utilized.

Pharmaceutical and nutraceutical compounds and activities

Antioxidative activity: Peptides derived from the hydrolysis of the skin gelatin with the enzyme pronase E displayed strong antioxidative activity on peroxidation of linoleic acid (Kim et al., 2001).

The oligopeptides of the hydrolysate derived from the skin of this species showed 2,2-diphenyl-1-picrylhydrazyl radical scavenging activity (DPPH). An IC_{50} value of 2.5 mg/mL was recorded for DPPH, and reducing power was 0.14 at 1 mg/mL (Jia et al., 2010).

The peptide APH-V with the amino acid sequence of Leu–Pro–His-Ser–Gly–Tyr derived from the frame the protein hydrolysate of this species showed the highest antioxidative activity with 35% scavenging on hydroxyl radical at 53.6 µM (Je et al., 2005).

ACE-inhibitory activity: The peptides derived from the protein hydrolysates of the backbone, frame, and skin of this species displayed—ACE-inhibitory activity (Wang et al., 2017).

Je et al. (2004) reported that the derived from frame protein hydrolysate of this species showed ACE-inhibitory peptide with an IC_{50} value of 14.7 µM.

Gadus macrocephalus Tilesius, 1810

Source: NOAA FishWatch. Public domain.

Common name(s): Pacific cod

Global distribution: Boreal; Atlantic Ocean: Chukchi and Beaufort Seas eastward across Arctic Canada to western Greenland; south in the Atlantic to Gulf of St. Lawrence. Pacific Ocean: eastern Pacific to southern California at Santa Monica, and in the western Pacific to the Yellow Sea

Ecology/Habitat: This demersal, oceanodromous species lives in both marine and brackish water areas at a depth range of 10–1280 m. It is found mainly along the continental shelf and upper slopes forming schools.

Maximum length and weight: 119 cm; 22.7 kg

Food and feeding: Adults of this species feed on fishes, octopi, crustaceans, and worms. However, young feeds mainly on copepods and other crustaceans.

Uses: Highly commercial fisheries exist for this species. It is also a game fish. It is marketed fresh and frozen for human consumption, and also dried or salted and smoked. It is used in Chinese medicine.

Pharmaceutical and nutraceutical compounds and activities

Antioxidative and ACE-inhibitory activity: The papain hydrolysate from the skin gelatin of this species showed significant antioxidant activity (75% at 500 g/mL). Further, two peptides with the amino acid sequence of Thr–Cys–Ser–Pro (388 Da) and Thr–Gly–Gly–Gly–Asn–Val, respectively, from the above hydrolysate showed potential ACE inhibitory activity (Ngo et al., 2011).

Iron-chelating activity: Three peptides with the amino acid sequence GPAG-PHGPPGKDGR, AGPHGPPGKDGR, and AGPAGPAGAR, respectively, derived from the skin gelatin of this species exhibited iron-chelating activity with high affinity to ferrous ions. This finding suggests the potential application of gelatin-derived peptides to combat iron deficiency (Wu et al., 2017).

Gadus morhua Linnaeus, 1758

Source: Hans-Petter Fjeld, https://creativecommons.org/licenses/by-sa/2.5/deed.en

Common name(s): Atlantic cod

Global distribution: Temperate; Western Atlantic Ocean: Cape Hatteras, North Carolina, Greenland, and Labrador Sea; Eastern Atlantic Ocean: Baltic Sea, North Sea, Iceland, and the Barents Sea

Habitat: This benthopelagic species dwells in marine and brackish water areas at a depth range of 0–600 m. It is oceanodromous forming schools during day.

Maximum length and weight: 200 cm: 96.0 kg

Food and feeding: It is an omnivorous feeder on invertebrates and fish.

Uses: This species has high commercial fisheries with aquaculture and game fish value. It is sold fresh, dried, salted, smoked, and frozen.

Pharmaceutical and nutraceutical compounds and activities

Antiproliferative activity: The peptides derived from the hydrolysate of this species showed growth inhibition of MCF-7/6 and MDA-MB-231, human breast cancer cell lines (Picot et al., 2006).

Antioxidative anticancer activity: The peptides derived from the backbone of this fish displayed DPPH scavenging activity (Nurdiani et al., 2017).

Nutraceutical and bioactive properties: The liver oil of this species has been reported to possess both nutraceutical and bioactive properties (antibacterial, antifungal, and antiviral) owing to its rich omega-3 fatty acids, namely, EPA and DHA, and vitamins A and D (Sego, 2017).

Micromesistius poutassou (Risso, 1827)

Source: H. Gervais. Public domain.

Common name(s): Blue whiting

Global distribution: North Atlantic: from western Barents Sea; Spitzbergen, Iceland and Greenland, Skagerrak; Kattegat to Morocco; western Mediterranean and western North Atlantic.

Habitat: This mesopelagic species occurring at aq. depth range of 30–3000 m.

Maximum length and weight: 55.5 cm; 830.00 g

Food and feeding: Its chief food items include small crustaceans such as euphausiids and amphipods, and fishes occasionally.

Uses: Highly commercial fisheries exist for this species. While a small catch is sold fresh and frozen, a large portion of the catch is used oil and fishmeal production.

Pharmaceutical and nutraceutical compounds and activities

Antiproliferative activity: The peptides derived from the hydrolysate of this species showed growth inhibition of human breast cancer cell lines, namely,

MCF-7/6 cell line and MDA-MB-231 cell line and percentage inhibition values were 22.3–26.3% and 13.5–29.8%, respectively (Picot et al., 2006).

Antioxidant and ACE-inhibitory properties: Protein hydrolysates prepared using subtilisin showed potent antioxidant and ACE-inhibitory properties (Moreno et al., 2017).

Others: Cudennec et al. (2012) reported that the consumption of muscle hydrolysate of this species led to a decrease in the body-weight gain.

Pollachius pollachius **(Linnaeus, 1758)**

Source: © Citron. https://creativecommons.org/licenses/by-sa/3.0/

Order: Gadiformes; Family: Gadidae

Common name(s): European Pollock

Global distribution: Temperate; Northeastern Atlantic: Norway, Faeroes, Iceland, and Bay of Biscay

Habitat: This pelagic to benthopelagic species is often found at hard rocky areas with a depth range of 40–200 m. Adults are oceanodromous forming dense shoals in spawning grounds. Juveniles are pelagic and are often found in rocky areas, kelp beds, sandy shores, and estuaries.

Max length and weight: 130 cm; 18.1 kg

Food and feeding: While adult fishes prey chiefly on other fishes, juveniles feed on small crustaceans, such as copepods, amphipods, and euphausiids, and small fish.

Uses: This species has commercial fisheries and is sold fresh and frozen.

Pharmaceutical and nutraceutical compounds and activities

Cytotoxicity: The peptides derived from the fresh filleting by-products of this species showed cytotoxicity against breast cancer cell lines (Nurdiani et al., 2017).

Nutraceutical property: Diet containing the protein hydrolysate at 230 g/kg for 26 days in rats reduced liver and visceral lipids (Rabiei et al., 2017).

Pollachius virens **(Linnaeus, 1758)**

Source: Tino Strauss. https://creativecommons.org/licenses/by-sa/2.0/deed.en

Common name(s): Saithe

Global distribution: Temperate; Western Atlantic: from Hudson Strait to North Carolina; Barents Sea; Spitsbergen to Bay of Biscay; Iceland and Greenland

Habitat: This demersal, gregarious species is found both in inshore and offshore waters at a depth range of 37–364 m. It is oceanodromous entering coastal waters in spring and returning to deeper waters in winter.

Maximum length and weight: 130 cm, 32.0 kg

Food and feeding: Adults prey chiefly upon other fishes and juveniles feed largely on small crustaceans such as copepods, amphipods, and euphausiids.

Uses: This species has high commercial fisheries and is sold fresh and frozen.

Pharmaceutical and nutraceutical compounds and activities

ACE-inhibitory activity: The peptides derived from the ultrafiltrated protein hydrolysates of this species showed ACE-inhibitory activity and the recorded values are given below:

ACE-inhibitory activity of protein hydrolysates of *P. virens*

Sample	Protein concentration (mg/mL)	ACE inhibition (%)	IC_{50} (mg/mL)
5 kDa	14.6	95.2	1.1
10 kDa	14.9	86.0	1.2
30 kDa	10.8	51.2	11.0

Source: Hamaguchi et al. (2008).

2.1.11   LINGS (ORDER: GADIFORMES; FAMILY: LOTIDAE)

Molva molva (Linnaeus, 1758)

Source: Image by Krüger. Public domain.

Common name(s): White fish, common ling

Global distribution: Temperate; Northwest Atlantic: Barents Sean, Iceland, Morocco, Mediterranean, southern Greenland, and Canada

Habitat: This demersal, oceanodromous species occurs mainly on rocky bottoms at a depth range of 100–1000 m.

Maximum length and weight: 200 cm; 45.0 kg

Food and feeding: Its food items include fish like cod, herring and flatfish; lobsters, cephalopods, and starfishes.

Uses: This species has high commercial fisheries and is sold fresh, dried, salted, or frozen. It is also a game fish and is occasionally used for fishmeal preparation.

Pharmaceutical and nutraceutical compounds and activities

Antioxidant activity: The peptides derived from the hydrolyzed skin of this species have shown antioxidant activity by reducing the oxidative stress (Pangestuti and Kim, 2017).

Nutraceutical properties: The product Fortidium, and its autolysate commercially known as Fortidium, is believed to possess bioactivities like reduction of oxidative stress and lowering of glycemic index (Hernández-Ledesma and Herrero, 2013).

2.1.12   HAKES (ORDER: GADIFORMES; FAMILY: MERLUCCIIDAE)

Merluccius productus (Ayres, 1855)

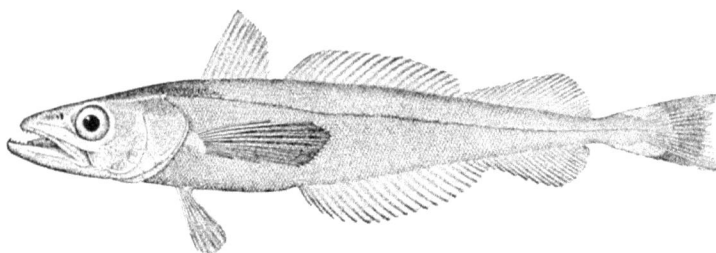

Source: Image by Anonymous. Bulletin of the United States Fish Commission, Public domain

Common name(s): North Pacific hake

Global distribution: Temperate; eastern Pacific: west coast of North America (Vancouver Island to Gulf of California)

Habitat: Marine; brackish; this pelagic species living in neritic habitats at a depth range of 0–1000 m. Its occurrence in oceanic and brackish water areas has also been reported. Adults often form large schools in waters overlying the continental shelf and slope.

Maximum length and weight: 91.0 cm; 1.2 kg

Food and feeding: It is a nocturnal feeder on other fishes and invertebrates.

Uses: It has high commercial fisheries and is sold fresh and frozen. It is also used for fishmeal production.

Pharmaceutical and nutraceutical compounds and activities

ACE-inhibitory activity: The hydrolysates of this species showed ACE-inhibitory activity with an IC$_{50}$ value of 44 µg/mL (Cinq-Mars and Li-Chan, 2007).

Immunomodulatory effects: The peptides of the protein hydrolysates of this species showed Immunomodulatory effects (Wang et al., 2017).

2.1.13   SAND EELS (ORDER: GASTEROSTEIFORMES; FAMILY: HYPOPTYCHIDAE)

Hypoptychus dybowskii **Steindachner, 1880**

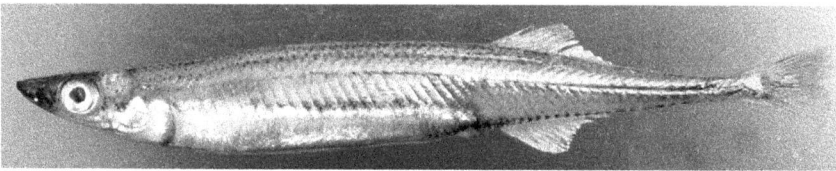

Source: Courtesy of IUCN Seahorse, Pipefish and Seadragon Specialist Group (SPS SG)

Common name(s): Korean sandlance

Global distribution: Temperate; north Pacific: Sakhalin, Russia to northern Japan

Habitat: It is a benthopelagic species living in shallow, coastal waters

Maximum length: 10.0 cm

Food and feeding: Its food items include mainly small invertebrates and fish larvae.

Uses: Commercial fisheries exist for this species.

Pharmaceutical and nutraceutical compounds and activities

Antioxidant activity: The papain hydrolysate from the protein of this fish and purified peptides showed antioxidant activity. While the DPPH radical scavenging activity of the papain hydrolysate was 77.4% at 1.0 mg/mL, the purified peptide showed an inhibitory effect against DNA oxidation induced by hydroxyl radical with an EC50 value of 22.75 µM (Lee et al., 2011; Venkatesan et al., 2017).

2.1.14   MILKFISH (ORDER: GONORYNCHIFORMES; FAMILY: CHANIDAE)

Chanos chanos (Forsskål, 1775)

Common name(s): Milkfish

Global distribution: Tropical; Pacific and Indian oceans; from the eastern coast of Africa to the western coast of the Americas

Habitat: It is a benthopelagic fish occurring in marine, freshwater, and brackish water areas at a depth range of 1–30 m. It is an amphidromous species entering both estuaries and freshwater streams and is forming small to large schools near the coasts and islands.

Maximum length and weight: 180 cm; 14.0 kg

Food and feeding: Adults fishes feed on algae, small benthic invertebrates, and fish eggs and larvae. Larvae of this fish, however, feed on zooplankton.

Uses: This species has high commercial fisheries. It has also aquaculture, aquarium, and game fish value.

Pharmaceutical and nutraceutical compounds and activities

Antioxidant, anti-inflammatory, and DNA-protective activities: The collagen peptides derived from the scales of this species showed high antioxidant activity, and anti-inflammatory activity by reducing lipoxygenase activity and nitric oxide (NO·) radicals. Further, these peptides have been reported to protect against cyclobutane di-pyrimidine production and DNA single-strand breaks, which are responsible for harmful UV radiation and H_2O_2 (Chen et al., 2018).

2.1.15 MULLETS OR GRAY MULLETS (ORDER: MUGILIFORMES; FAMILY: MUGILIDAE)

Chelon auratus (Risso, 1810) (=*Liza aurata*)

Common name(s): Golden gray mullet

Global distribution: Temperate; Eastern Atlantic: Scotland to Cape Verde; Mediterranean and Black Sea; from southern Norway to Morocco; rare off Mauritania

Habitat: This neritic, catadromous species lives in marine and brackish water areas; depth 10 m; adults are usually in schools, entering lagoons and lower estuaries rarely entering freshwater.

Maximum length: 59.0 cm

Food and feeding: Its chief food items include small benthic organisms and detritus, and occasionally on insects. Juveniles, however, feed only on zooplankton.

Uses: This species has commercial fisheries with aquaculture value.

Pharmaceutical and nutraceutical compounds and activities

Antioxidative properties: The protein hydrolysates of this species showed DPPH scavenging activity which was found to increase with increased hydrolysis. When the degrees of hydrolysis increased from 12.4% to 15.4%, the DPPH radical scavenging activity also increased from 50.1% to 72% (Bkhairia et al., 2029).

Anti-inflammatory (anti-5-lipoxygenase (5-LOX)) activity: The head protein hydrolysate treated with the different concentrations (D1, D2, and D3) of neutrase for estimating this activity. Among the different concentrations, DH3 exhibited the highest anti-inflammatory activity (57.4%) in comparison with DH2 (46.3%) and DH1 (35.0%), at the equal concentration. However, the undigested head protein of this species did not show any inhibition of the 5-LOX enzyme activity at the same concentration (Bkhairia et al., 2019).

Mugil cephalus Linnaeus, 1758

Common name(s): Flathead gray mullet

Global distribution: Cosmopolitan in tropical and subtropical seas; Atlantic: from Bay of Biscay southward; Mediterranean and Black Sea

Habitat: This catadromous species is found in coastal waters, estuaries, and freshwater environments at a depth range of 0–10 m. Adults are highly euryhaline and are capable of tolerating even zero salinity.

Maximum length: 100.0 cm

Food and feeding: This is a diurnal feeder and its food items include mainly zooplankton, dead plant matter, and detritus.

Uses: This species has high commercial fisheries with aquaculture and game fish values. It is also used as fish bait.

Pharmaceutical and nutraceutical compounds and activities

Anticancer activity: The mucus of this species showed anticancer activity against laryngeal cancer cell lines (Balasubramanian et al., 2016). The percentage death of cancer cells at different concentration of the mucus are given in the following table.

Anticancer activity of mucus of *M. cephalus* on laryngeal cancer cell lines.

Concentration (µg/mL)	Cell viability (%)
1000	86.3
500	80.9
250	74.4
125	70.2

Source: Balasubramanian et al. (2016).

Hemolytic activity: The epidermal mucus of this species exhibited hemolytic activity on goat RBC and chicken RBC (Balasubramanian et al., 2016).

Antibacterial activity: The muscle tissue and proteins of this species showed in-vitro antimicrobial activity against *E. coli, Proteus mirabilis, S. aureus, P. aeruginosa*, and *Klebsiella pneumoniae*. A maximum inhibition zone of 29 mm was observed against *P. mirabilis* followed by *S. aureus* with an inhibition zone of 16 mm. *P. aeruginosa* showed minimum (4.14 mm) inhibitory activity (Deivasigamani et al., 2017).

2.1.16 UNICORNFISHES (ORDER: PERCIFORMES; FAMILY: ACANTHURIDAE)

Naso thynnoides (Cuvier, 1829)

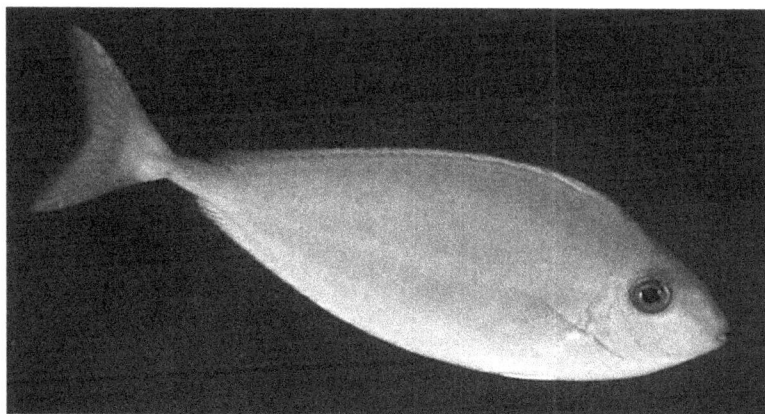

Source: Image by Izuzuki. https://creativecommons.org/licenses/by-sa/2.5/deed.en

Common name(s): Oneknife unicornfish

Global distribution: Tropical; Pacific and Indian Oceans

Habitat: This is commonly found in rocky areas and coral reef at a depth range of 2–40 m. It may form large schools.

Maximum length: 40.0 cm

Food and feeding: Feeds on zooplankton and algae

Uses: Commercial fisheries exist for this species. It is also a commercially important aquarium fish.

Pharmaceutical and nutraceutical compounds and activities

Antioxidant, ACE-inhibitory, and functional properties: The hydro-lysate derived from the skin gelatin of this fish showed DPPH scavenging activity and ferric reducing antioxidant power with the values of 63% and 25.90 Trolox equivalent (mM/mg), respectively. Further, this hydrolysate showed ACE-inhibitory activity with 33.97% and an IC_{50} value of 10.17 µg/mL. Furthermore, these hydrolysates may also be of great use as food ingredients owing to their rich glycine (40.70%) and glutamic acid + glutamine (25.40%) contents (Alolod et al., 2019).

2.1.17   CLIMBING PERCHES (ORDER: PERCIFORMES; FAMILY: ANABANTIDAE)

Anabas testudineus (Bloch, 1792)

Common name(s): Climbing perch

Global distribution: Tropical; Asia: China, India, Indonesia, Laos, Malaysia, Myanmar, Nepal, Pakistan, Singapore, Sri Lanka, Taiwan, Thailand, Vietnam, Andaman Island, Bangladesh, and Cambodia

Habitat: This demersal species is found in freshwater habitats such as canals, lakes, ponds, and swamps; and estuaries. It is also a potamodromous species

Maximum length and weight: 25.0 cm; 117 g

Food and feeding: Feed on macrophytic vegetation, shrimps, and fish fry

Uses: It is a commercial species of fisheries, aquaculture, and aquarium.

Pharmaceutical and nutraceutical compounds and activities

Antibacterial activity: The epidermal mucus extracts of this species showed significant activity against *P. aeruginosa*. Least activity was however observed against methicillin-resistant *Staphylococcus aureus* (Al-Rasheed et al., 2018).

Others: This species has been reported to possess anti-inflammatory, antinociceptive, antimicrobial, and anticancer properties (Rajani and Alka, 2015).

2.1.18   BLENNIES (ORDER: PERCIFORMES; FAMILY: BLENNIIDAE)

Salaria basilisca (Valenciennes, 1836)

Source: Image by Alberto Piras. Public domain.

Common name(s): Zebra blenny

Global distribution: Subtropical; Mediterranean Sea: Adriatic Sea, Tunisia, and Turkey

Habitat: This demersal species is commonly seen among seagrass and occasionally on rocky bottoms at a depth range of 2–15 m.

Maximum length: 18.0 cm

Food and feeding: Not reported

Uses: Not reported.

Pharmaceutical and nutraceutical compounds and activities

Antioxidant properties and ACE-inhibitory activities: The protein hydrolysates of this species have shown antioxidant and ACE-inhibitory activity in alloxan-induced diabetic rats. Further, it is also reported that these hydrolysates protected the kidney function efficiently in these experimental rats, as evidenced by a significant reduction in the creatinine, uric acid, and urea contents (Ktari et al., 2012, 2014).

Hypoglycemic and hypolipidemic properties: The protein hydrolysates of the muscle of this species have shown hypoglycemic and hypolipidemic effects in alloxan-induced diabetic rats. This is largely due to the rich leucine content of these hydrolysates. Leucine is believed to be one of the active ingredients for blood glucose control by inducing insulin release in both rats and humans. Further, these hydrolysates decreased significantly the triglyceride (TG), total cholesterol (TC), and LDL-cholesterol (LDL-c) levels in the serum and liver of diabetic rats; and increased the HDL-cholesterol level. Furthermore, these hydrolysates displayed potent protective effects against heart attack markers by reversing myocardial enzyme serum back to normal levels (Ktari et al., 2013).

Hypocholesterolemic effects and antioxidative activities: Treatment with the protein hydrolysates of this species increased the level of high-density lipoprotein cholesterol and reduced the levels of triglycerides, total cholesterol, and low-density lipoprotein cholesterol in experimental rats. Further, there was significant normalization of thiobarbituric acid-reactive substance levels as well as catalase, superoxide dismutase, and glutathione peroxidase activities in renal and hepatic tissues. Furthermore, significant protective effects were also recorded in the liver and kidney functions which were associated with a marked reduction in the level of serum urea, uric acid, creatinine, alkaline phosphatase, and alanine aminotransferase (Ktari et al., 2015; Rabiei et al., 2017).

2.1.19 JACKS, POMPANOS, AND SCADS (ORDER: PERCIFORMES; FAMILY: CARANGIDAE)

Caranx ignobilis (Forsskål, 1775)

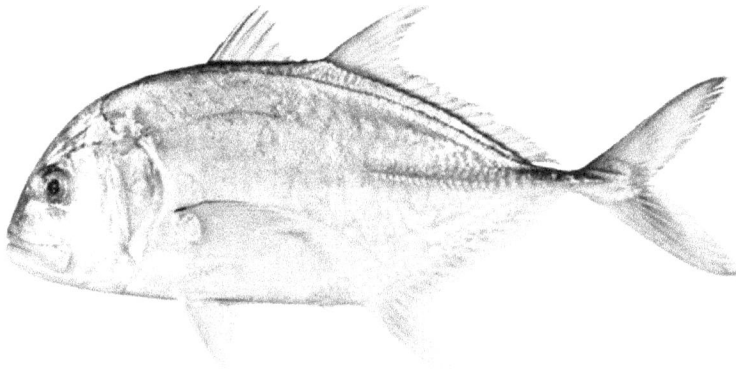

Common name(s): Giant trevally

Global distribution: Tropical; Indo-Pacific: from South Africa in the west to Hawaii in the east

Habitat: This species occurs in sand and rock habitats of marine and brackish water areas at a depth range of 10–188 m.

Maximum length and weight: 170 cm; 80.0 kg

Food and feeding: Its food items include crustaceans, such as crabs and spiny lobsters, and fishes. It is a nocturnal feeder.

Uses: It is known for its commercial fisheries and aquaculture. It is also a game fish with aquarium value.

Pharmaceutical and nutraceutical compounds and activities

Antioxidant properties: The two fraction of protein hydrolysates of this species showed a high level of essential amino acids including the most abundant amino acids, namely, His (12.01% and 7.08%), Phe (7.05% and 6.18%), and Lys (6.76% and 4.74%). These fractions showed antioxidant properties in terms of effective inhibition of lipid peroxidation (Nazeer and Kulandai, 2012).

Caranx melampygus **Cuvier, 1833**

Courtesy of: Wikimedia

Common name(s): Bluefin trevally

Global distribution: Tropical; Indian and Pacific Oceans: from Eastern Africa to Central America including Japan and Australia

Habitat: This pelagic species is found in reefs of coastal and oceanic waters at a depth range of 0–190 m. It may form schools also. Juveniles are entering shallow inshore waters and rivers seasonally.

Maximum length and weight: 117 cm; 43.5 kg

Food and feeding: Adults feed mainly on other fishes and crustaceans

Uses: This species has commercial fisheries with aquaculture, aquarium, and game fish values.

Pharmaceutical and nutraceutical compounds and activities

Wound healing potential: The collagen film prepared from the bones of this fish possessed better mechanical properties. The in-vitro studies demonstrated its biocompatible nature. Acceleration of wound healing in collage film-treated rats was evident in the in-vivo studies. As a wound dressing material, this collagen film is also very useful (Rethinam et al., 2016).

Decapterus maruadsi (Temminck & Schlegel, 1843)

Common name(s): Blue mackerel scad, Japanese scad, and round scad

Global distribution: Tropical; western central Pacific and Eastern Indian Ocean

Habitat: This reef-associated species is found in warm coastal waters at a depth range of 0–20 m. Adults enter semi-enclosed seas.

Maximum length: 35.0 cm

Food and feeding: It feeds mainly on planktonic crustaceans such as euphausiids and copepods, and small fish.

Uses: It is a commercially important species in China where it is sold fresh and dried-salted. It is also used in the preparation of the nutritionally important surimi (ground meat).

Pharmaceutical and nutraceutical compounds and activities

Antioxidative activity: The high protein content (15.5%) and low lipid content (1.8%) of the muscle of this species suggest that this fish may serve as a functional food (Jiang et al., 2014). Further, the hydrolysate of this fish prepared with alcalase showed high antioxidative activity. The DPPH scavenging activities of its different hydrolysates are given in the following table.

Antioxidative activities of *D. maruadsi* protein hydrolysates prepared with different enzymes.

Hydrolysate	DPPH scavenging activity (%)
Trypsin	39.4
Papain	40.2
Alcalase	39.4
Neutral	32.3
Pepsin	32.6

Source: Jiang et al. (2014).

Thiansilakul et al. (2007) reported that the hydrolysate prepared with flavorzyme exhibited a higher DPPH radical scavenging activity and reducing power

Magalapsis cordyla (Linnaeus, 1758)

Common name(s): Finletted mackerel scad, cordyla scad, Torpedo scad, horse mackerel

Global distribution: Tropical; Indo-Western Indian Ocean; western Pacific Ocean: from Japan to Australia

Habitat: It is a pelagic species living in inshore waters of the continental shelf and is forming schools.

Maximum length and weight: 80 cm; 4 kg

Food and feeding: It is a carnivorous species feeding chiefly on *Acetes indicus* and other fishes like *Stolephorus* species; on juveniles of *Coilia*, *Sardinella*, *Nemipterus*, *Thryssa*, *Trichiurus*, *Apogon*, and sciaenids; and on young ones of molluscs and ostracods.

Uses: This species has high commercial fisheries and is sold fresh and dried-salted.

Pharmaceutical and nutraceutical compounds and activities

Antioxidant activity: The peptide with the amino acid sequence Ala–Cys–Phe–Le derived from the visceral protein of this species showed activity against DPPH and hydroxyl radicals with percentage values of 89.2 and 59.1, respectively. Further, this peptide demonstrated high activity against polyunsaturated fatty acid (PUFA) peroxidation (Kumar et al., 2011; Nurdiani et al., 2017).

Wound healing properties: The collagen from the bone of this species has been reported to heal significantly the excision wounds made on Wistar rats. This peptide has excellent medicinal importance and reveals a possible way to convert this waste into a medicinally important source (Kumar et al., 2012).

Parastromateus niger (Bloch, 1795)

Common name(s): Black pomfret

Global distribution: Tropical: Indo-West Pacific region

Habitat: This pelagic species occurs in the muddy bottoms of the continental shelf at a depth range of 15–40 m; it often forms large schools and enters estuaries.

Maximum length and weight: 75.0 cm

Food and feeding: It feeds mainly on zooplankton.

Uses: This species has very high commercial fisheries. It is delicious and is mostly marketed fresh.

Pharmaceutical and nutraceutical compounds and activities

Antioxidant activities: The crude visceral protein hydrolysate of this species exhibited antioxidant activities. At a concentration of 1 mg/mL, the activity was with 54% (Ganesh et al., 2011).

Selaroides leptolepis (Cuvier, 1833)

Common name(s): Yellowstripe scad, yellow-striped trevally

Global distribution: Tropical; Western Indian Ocean: Sri Lanka; Indo-West Pacific: Okinawa, Japan, the Philippines, Indonesia, Australia, Bay of Bengal, and Gulf of Thailand

Habitat: This reef-associated species occurs in inshore waters (less than 50 m) of the continental shelf. Adults which are amphidromous may also be seen in and brackish water areas.

Maximum length and weight: 22 cm; 625 g

Food and feeding: It feeds on euphausiid, ostracods, gastropods, and small fish.

Uses: Commercial fisheries exist for this species.

Pharmaceutical and nutraceutical compounds and activities

Antioxidant activities: The protein hydrolysates derived from this species using the enzymes, alcalase and flavorzyme demonstrated significant antioxidant activity (Khora, 2013).

Nutraceutical properties: This medium-fat species (28%) with a high content of polyunsaturated fatty acids, namely, DHA (20%) and EPA (12.6%), has industrial applications especially in the extraction of high-value biomolecules of nutraceutical value (Viet and Ohshima, 2014).

Seriola lalandi **Valenciennes, 1833**

Common name(s): Yellowtail kingfish, yellowtail amberjack, greatamber jack, California yellowtail

Global distribution: Subtropical; Northwest Pacific, South Pacific, and Northeast Pacific

Habitat: Marine; brackish; this benthopelagic species lives both in marine and brackish environments at a depth range of 3–825 m. Adults are solitary or in small groups and are commonly found in the vicinity of rocky shores, coral reefs, and islands.

Maximum length and weight: 250 cm; 96.8 kg

Food and feeding: Adults feed on small fish, squid, and crustaceans.

Uses: This species has minor commercial fisheries and is sold fresh and salted dried. It may have game fish and aquaculture values.

Pharmaceutical and nutraceutical compounds and activities

Antimicrobial activity: Two peptides, namely, piscidin and hepcidin, derived from the protein hydrolysate of this fish showed antibacterial activity (Muncaster et al., 2018).

Trachurus japonicus (Temminck & Schlegel, 1844)

Common name(s): Horse mackerel, Japanese jack mackerel

Global distribution: Tropical; Western North Pacific: Japan

Habitat: This pelagic species is found in neritic zone at a depth range of 0–275 m. Adults are oceanodromous.

Maximum length and weight: 50.0 cm; 660 g

Food and feeding: The chief food items of this species include copepods, shrimps, and other small fishes.

Uses: This species has high commercial fisheries. It is also used in commercial aquaculture. It is utilized canned for human consumption and is also made into fishmeal.

Pharmaceutical and nutraceutical compounds and activities

Antioxidant properties: The fermented fish paste of this species has shown displayed DPPH radical scavenging activity, and inhibition of hydroxyl radicals and NO. The inhibition of hydroxyl radicals by above paste extracts indicated that the substrate responsible for hydroxyl radical scavenging was rapidly developed during the early stages of fermentation. In the advanced stages of fermentation (over 135 days) for fish miso stored at 25°C, the scavenging activity mostly remained stable. The inhibition of NO by fish miso extracts revealed that under the experimental conditions used, radical scavenging activity rapidly increased during a fermentation period of 90 days until all NO radicals were scavenged. These findings suggest that the fish miso fermented using soybean koji inoculated with *A. oryzae*, bears good flavor and this could be used as a functional food (Giri et al., 2012).

2.1.20 TREVALLAS (ORDER: PERCIFORMES; FAMILY: CENTROLOPHIDAE)

Seriolella punctata (Forster, 1801)

Common name(s): Silver warehou

Global distribution: Temperate; southern Pacific and Southern Indian oceans

Habitat: This benthopelagic and oceanodromous species is found in continental shelf and slope waters at a depth range 27–650 m. Adults are rarely entering bays and brackish water areas.

Maximum length: 66.0 cm

Food and feeding: Adults of this species mainly eat planktonic tunicates.

Uses: Commercial fisheries exist for this species.

Pharmaceutical and nutraceutical compounds and activities

Antioxidant and ACE-inhibitory activity: The peptides derived from the protein hydrolysates of this species showed DPPH radical scavenging activity. Further, its protein hydrolysate treated with acid fungal protease showed ACE-inhibitory activity (11.58%) (Nurdiani et al., 2016).

2.1.21   SNAKEHEADS (ORDER: PERCIFORMES; FAMILY: CHANNIDAE)

Channa striatus (Bloch, 1793)

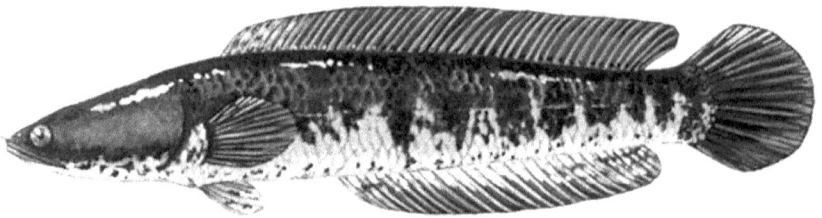

Common name(s): Striped snakehead

Global distribution: Asia: Pakistan, Sri Lanka, Indonesia, the Philippines, Thailand, Cambodia, and China

Habitat: This benthopelagic, potamodromous species is found in sluggish or standing waters of both freshwater and brackish water areas at a depth range of 1–10 m. Adults are commonly seen in ponds, streams, rivers, and swamps. Maximum length and weight: 100.0 cm; 3.0 kg

Food and feeding: Its food items include a wide variety of animals such as earthworms, tadpoles, crustaceans, fish, frogs, snakes, insects, and so on.

Uses: This fish is highly commercial fisheries with aquaculture and aquarium values.

Pharmaceutical and nutraceutical compounds and activities

Anti-inflammatory antipyretic properties: In experimental rats, the aqueous extracts of this species showed anti-inflammatory activity by reducing the soft tissue swelling and synovial inflammation. Therefore, this fish may be useful in the treatment of joint diseases such as rheumatoid arthritis (Shafri and Manan, 2012). Zakaria et al. (2008) reported that the antipyretic activity of the aqueous extract of this species is largely due to its anti-inflammatory property.

Antibacterial activity: The ethanol extract of this species showed activity against *Staphylococcus aureus* (Mat Jais et al., 2008). Wei et al. (2010) reported that its acidic extract acted against *Klebsiella pneumoniae*,

Pseudomonas aeruginosa, and *B. subtilis*. The mucus extracts of its skin and intestine of this species on the other hand showed activity against *Aeromonas hydrophila* and *Pseudomonas aeruginosa* (Shafri and Manan, 2012).

Antifungal activities: The ethanolic extract of the fillets of this species displayed antifungal activity against *Aleurisma keratinophilum*, *Cordyceps militaris*, *Botrytis pyramidalis*, *Neurospora crassa*, and *Paecilomyces fumosaroseus* (Jais et al., 2008).

Cardiological effects: The skin extract of this species has been found to contain potent active compound, The cardiotoxic factor II (CTF-II) found in the skin extract of this species has been reported to show cardiological effects (Karmakar et al., 2002).

ACE-inhibitory activity: Shafri and Manan (2012) reported that this species possesses ACE-inhibitory activity which make this species as a functional food and preventative medicine for hypertensive patients (Shafri and Manan, 2012).

2.1.22   CICHLIDS (ORDER: PERCIFORMES; FAMILY: CICHLIDAE)

***Oreochromis mossambicus* (Peters, 1852)**

Courtesy of: Wikimedia

Common name(s): Mozambique tilapia
Global distribution: Tropical; Africa

Habitat: This benthopelagic and amphidromous species inhabits reservoirs, rivers, creeks, drains, swamps and tidal creeks; and commonly over mud bottoms, often in well-vegetated areas. It is also found in warm weedy pools of sluggish streams, canals, ponds, estuaries, and coastal lakes at a depth range of 1–12 m.

Maximum length and weight: 39.0 cm; 1.1 kg

Food and feeding: This omnivorous species feeds mainly on algae and phytoplankton but also takes zooplankton, small insects, shrimps, earthworms, and aquatic macrophytes. Adults may also take detritus and small fishes, and they are occasionally cannibalistic on their own young. Juveniles are, however, carnivorous or omnivorous.

Uses: This species has high commercial fisheries and it also possesses aquaculture and aquarium values.

Pharmaceutical and nutraceutical compounds and activities

Anticancer and antimicrobial activities: The hepcidin TH2–3 a synthetic 20-mer peptide of this species has been reported to inhibit the proliferation and migration of human fibrosarcoma cell (HT1080 cell line); and antimicrobial activity (Chang et al., 2011; Huang et al., 2007).

Oreochromis niloticus (Linnaeus, 1758)

Common name(s): Nile tilapia

Global distribution: It is a global invasive species and is native to Africa. Now it is found distributed in all countries within tropics.

Habitat: This diurnal species lives in a variety of freshwater areas such as rivers, lakes, sewage canals, and irrigation channels; and brackish water environments. However, it is dominant in shallow waters.

Maximum length and weight: 60.0 cm; 4.3 kg

Food and feeding: It is an omnivore feeding on a variety of food items such as phytoplankton, plants, small invertebrates, bottom fauna, and detritus.

Uses: This species has high commercial fisheries and it has aquaculture value. It is sold fresh and frozen.

Pharmaceutical and nutraceutical compounds and activities

ACE-inhibitory activity: The hydrolysate of the collagen extracted from the skin of this species showed an ACE-inhibitory activity with an IC_{50} value of 1.2 mg/mL. Further, the peptides purified from its hydrolysates had similar activity with a value of 0.8 mg/mL (Thuan-thong et al., 2017).

Antioxidant and cytotoxic activity: The alcalase-derived hydrolysate from the scales of this species showed significant antioxidant activity. Further, its peptide with the amino acid sequence Asp–Pro–Ala–Leu–Ala–Thr–Glu–Pro–Asp–Pro–Met–Pro–Phe has been reported to scavenge hydroxyl, DPPH, and superoxide radicals at the IC_{50} values of 7.56, 8.82, and 17.83 M, respectively. Furthermore, this peptide showed no cytotoxic properties on mouse macrophages (RAW 264.7) and human lung fibroblasts (MRC-5) (Ngo et al., 2010).

Antihypertensive activity: The peptide derived from the alcalase hydrolysate of this species showed significant antihypertensive activity (Vo et al., 2011).

Wound healing property: The collagen peptides of this species have also been reported to enhance the process of wound healing in experimental rabbits (Hu et al., 2017).

Pelmatolapia mariae (Boulenger, 1899) (=*Tilapia mariae*)

Common name(s): Spotted tilapia

Global distribution: Tropical; it is native to West Africa; from Tabou River (Côte d'Ivoire) to the Kribi River (Cameroon)

Habitat: This demersal species lives in still or flowing waters in rocky or mud-bottom areas of freshwater and brackish water.

Maximum length and weight: 32.3 cm; 1.4 kg

Food and feeding: It is an herbivore feeding mainly on phytoplankton (diatom and blue-green algae) and higher plants, and it also eats insects and shrimps.

Uses: It is an important food and source of protein, and also an important aquarium species.

Pharmaceutical and nutraceutical compounds and activities

ACE-inhibitory activity: A peptide, MILLER isolated from this species displayed ACE-inhibitory activity with an IC_{50} value of 0.12 μM (Korczek et al., 2018).

Antioxidant and hypoglycemic properties: The collagen peptide derived from the skin of this species demonstrated ABTS and DPPH scavenging activity with the values of 1.48 and 7.97 mg/mL, respectively. Further, this peptide has also shown hypoglycemic properties (Zhang et al., 2016a).

2.1.23   GOBY (ORDER: PERCIFORMES; FAMILY: GOBIIDAE)

Periophthalmodon schlosseri (Pallas, 1770)

Common name(s): Giant mudskipper

Global distribution: Tropical; Malaysia; Myanmar; Palau; Papua New Guinea; Australia; Brunei Darussalam; Cambodia; India; Indonesia; Lao People's Democratic Republic; the Philippines; Singapore; Thailand; Timor-Leste; Vietnam

Habitat: This demersal species lives in a variety of habitats such as sea, estuaries, and rivers at a depth range of 0–2 m. It is an amphibious air-breather and amphidromous.

Maximum length and weight: 240 mm

Food and feeding: It is carnivorous preying upon fiddler crabs, insects, and worms.

Uses: Minor commercial fisheries exist for this species.

Pharmaceutical and nutraceutical compounds and activities

Antibacterial activity: The organic mucus extracts of this species showed activity against clinical bacterial strains and values of zone of inhibition recorded are given in the following table.

Antibacterial activity of the organic mucus extracts against bacterial pathogens.

Species	Zone of inhibition (mm dia.)
Escherichia coli	15
Staphylococcus aureus	12
Salmonella typhi	18

Species	Zone of inhibition (mm dia.)
Vibrio cholerae	14
Proteus mirabilis	12
Pseudomonas aeruginosa	16
Klebsiella pneumoniae	12
Bacillus anthracis	10

Source: Mahadevan et al. (2019).

Antifungal activity: The organic mucus extract of this species showed activity against fungal strains and the values of zone of inhibition recorded on the different species are given in the following table.

Antifungal activity of organic mucus extract on fungal pathogens.

Species	Zone of inhibition (mm dia.)
Aspergillus flavus	13
Mucor sp.	16
Candida albicans	15
Trichoderma longibrachiatum	14

Source: Mahadevan et al. (2019).

Hemolytic activity: The aqueous and organic mucus extracts of this fish showed hemolytic activity with A, B, AB, and O groups of human blood and chicken erythrocytes (Mahadevan et al., 2019).

Zosterisessor ophiocephalus (Pallas, 1814)

Common name(s): Grass goby

Global distribution: Northeast Atlantic: Black Sea, Mediterranean, and Sea of Azov

Habitat: It dwells on the muddy bottoms or in eel-grass meadows in estuaries and lagoons.

Maximum length: 25 cm

Food and feeding: It is a benthic feeder and its food items include worms and crustaceans.

Uses: This species possesses commercial fisheries.

Pharmaceutical and nutraceutical compounds and activities

Antioxidant activities: The protein hydrolysates derived from the muscle of this species with alkaline proteases from Bacillus mojavensis A21, *Bacillus subtilis* A1, *Bacillus pumilus* A26, and *Bacillus licheniformis* NH1 and alcalase showed antioxidant activities. However, the hydrolysate from A26 proteases displayed very significant DPPH radical scavenging activity, followed by NH1 and A21 hydrolysates (Nasri et al., 2012).

Antioxidant and antidiabetic activity: Administration of the protein hydrolysates of this species with high-fructose fed rats resulted antidiabetic activity with the reduction of serum glucose, liver glycogen, α-amylase activity, and serum MDA. Further, this treatment increased the activities of liver antioxidant enzymes of the experimental rats (Rabiei et al., 2017).

Antioxidative and ACE-inhibitory activities: The protein hydrolysates derived from the muscle of this species showed ACE-inhibitory activity. Further, these hydrolysates which are rich in amino acids, such as Gly and Thr, demonstrated significant metal chelating activity and DPPH free radical scavenging activity and inhibited linoleic acid peroxidation (Nasri et al., 2014).

Hypolipidemic, cardioprotective, and anticoagulant properties: In Wistar rats fed with high-fat and fructose diet (HFFD), the protein hydrolysates (GPH) of this species showed hypolipidemic activity by increasing pancreatic lipase activity. Further, it is also reported that the administration of undigested goby protein (UGP) and GPHs to HFFD-fed rats was found to lower the serum TC, TG, and LDL-c. All these treatments also demonstrated anticoagulant activity by significantly reducing the atherogenic index and coagulant factor levels (thrombin and prothrombin). Furthermore, the UGP and its hydrolysates showed cardioprotective potential by lowering the risk of atherogenic and coronary artery disease (Nasri et al., 2018).

2.1.24   GRUNTS (ORDER: PERCIFORMES; FAMILY: HAEMULIDAE)

Brachydeuterus auritus **(Valenciennes, 1832)**

Common name(s): Bigeye grunt

Global distribution: Tropical; Eastern Atlantic: west coast of Africa; from Mauritania to Angola; also reported from Morocco

Habitat: This benthopelagic and semipelagic species dwells in the inhabitants of sandy and muddy bottoms of coastal areas at a depth range of 10–100 m. It normally stays at the bottom in the daytime and moves to the open water at night. It rarely visits lagoons and estuaries during sexual maturation.

Maximum length: 30.0 cm

Food and feeding: Its chief food items include invertebrates and other small fishes.

Uses: This species has high commercial fisheries and is sold fresh and smoked. Discarded fish is made into fish powder or fishmeal.

Pharmaceutical and nutraceutical compounds and activities

Nutraceutical properties: Fish powder made from this fish is rich in protein (70.4 g/100 g) and in minerals such as iron (8.92 mg/100 g), phosphorus

(94 mg/100 g), and calcium (2387 mg/100 g). Owing to these factors, this product may be an ideal source of micronutrients and protein especially for the poor, vulnerable groups (Abbey et al., 2017).

2.1.25   SEABASSES (ORDER: PERCIFORMES; FAMILY: LATEOLABRACIDAE)

Lateolabrax japonicus (Cuvier, 1828)

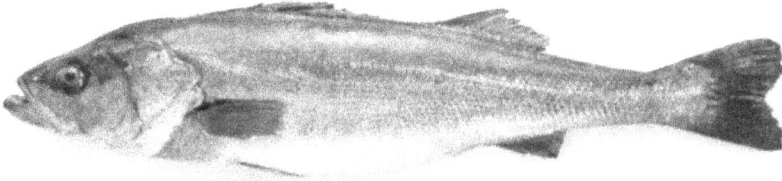

Common name(s): Asian seaperch, Japanese seabass

Global distribution: Subtropical; western Pacific: Japan to the South China Sea

Habitat: This species dwells in marine, freshwater, and brackish water habitats. It is also a catadromous fish.

Maximum length and weight: 102 cm; 8.7 kg

Food and feeding: It is predaceous, feeding on zooplankton at an early age and on small fish and shrimps as adults.

Uses: Commercial fisheries exist for this game fish. It is also a commercially important aquaculture: species. It is utilized as a food fish and in Chinese medicine.

Pharmaceutical and nutraceutical compounds and activities

Antimicrobial activity: Desriac et al. (https://www.academia.edu/13627658/Antimicrobial_Peptides_from_Fish) reported that its Hepcidin-like peptide demonstrated antimicrobial activity.

2.1.26   LATES PERCHES OR BARRAMUNDI (ORDER: PERCIFORMS; FAMILY: LATIDAE)

Lates calcarifer (Bloch, 1790)

Common name(s): Asian sea bass, barramundi

Global distribution: Tropical; Indo-West Pacific: Persian Gulf to China, Taiwan, and southern Japan; Papua New Guinea and Australia

Habitat: This demersal species occurs in a variety of habitats such as in freshwater, coastal waters, estuaries, and lagoons at a depth range of 10–40 m. It makes catadromous/diadromous migrations during spawning.

Maximum length and weight: 200 cm; 60.0 kg

Food and feeding: It feeds mainly on fishes and crustaceans.

Uses: This species has high commercial fisheries. It is also a commercially important aquaculture, aquarium, and game fish.

Pharmaceutical and nutraceutical compounds and activities

Antimicrobial and hemolytic activity: The skin and the head kidney of this species yielded moronecidin and dicentracin-like peptides which showed significant activity against Gram-positive bacteria. Further, among these peptides, dicentracin-like peptide showed more potent binding ability to all Gram-positive and Gram-negative bacteria. Furthermore, the moronecidin-like peptide showed potent hemolytic activity against human RBC (Taheri et al., 2018).

Antioxidative property: A total of four peptides with amino acid sequence Gly–Leu–Phe–Gly–Pro–Arg, Gly–Ala–Thr–Gly–Pro–Gln–Gly–Pro–Leu–Gly–Pro–Arg, Val–Leu–Gly–Pro–Phe, and Gln–Leu–Gly–Pro–Leu–Gly–Pro–Val have been isolated from the skin of this species. However, the peptide with Gly–Leu–Phe–Gly–Pro–Arg only showed significant anti-oxidant activity (81.41 mmol TE/μmol peptide) (Sae-Leaw et al., 2017). Nurdiani et al. (2017) also reported on the DPPH scavenging activity of its roe peptides.

2.1.27 EMPERORS (ORDER: PERCIFORMES; FAMILY: LETHRINIDAE)

Lethriuns atlanticus Valenciennes, 1830

Common name(s): Atlantic emperor

Global distribution: Tropical; Eastern Central Atlantic: Senegal to Gabon; Cape Verde, São Tome-Principe Islands, and Rolas Islands

Habitat: Marine; this reef-associated, nonmigratory species is found in shallow coastal waters at a depth range of 1–50 m.

Maximum length: 50.0 cm

Food and feeding: It feeds mainly on bottom-dwelling invertebrates such as crabs.

Uses: This species has commercial fisheries and the catch is sold fresh, smoked, and dried-salted.

Pharmaceutical and nutraceutical compounds and activities

Anticancer activity: The peptides derived from the hydrolysate of this species exhibited anticancer activity against MCF-7/6 and MDA-MB-231 cells (Picot et al., 2006).

2.1.28   SNAPPERS (ORDER: PERCIFORMES; FAMILY: LUTJANIDAE)

Lutjanus vitta (**Quoy & Gaimard, 1824**) (*=Lutjanus vita*)

Common name(s): Brownstripe red snapper

Global distribution: Temperate Australasia and northern Pacific and tropical Indo-Pacific

Habitat: This species occurs in the marine and brackish water areas at a depth range of 10–72 m. Adults are found associated with coral reefs, sponges, and sea whips. It may also form schools.

Maximum length: 40.0 cm

Food and feeding: Feeds on fishes, shrimps, crabs, and other benthic invertebrates.

Uses: Commercial fisheries exist for this species. It also has aquarium value.

Pharmaceutical and nutraceutical compounds and activities

Antioxidative and ACE-inhibitory activities: The two types of protein hydrolysates derived from the muscle of this species through 2-h and 1-h hydrolysis with pyloric ceca protease yielded antioxidative and

ACE-inhibitory activities. While 2-h hydrolysates displayed significant DPPH and ABTS radical scavenging activity and ferric reducing antioxidant power, 1-h hydrolysates showed higher ferrous-chelating activity and ACE-inhibitory activity (Khantaphant et al., 2011).

2.1.29 TEMPERATE BASSES (ORDER: PERCIFORMES; FAMILY: MORONIDAE)

Dicentrarchus labrax (Linnaeus, 1758)

Source: Image by Hans Hillewaert. https://creativecommons.org/licenses/by-sa/4.0/deed.en

Common name(s): European seabass

Global distribution: Subtropical; eastern Atlantic: Norway to Morocco; Canary Islands and Senegal; Mediterranean and Black Sea

Habitat: It is found in the benthic zones of marine, freshwater, and brackish water areas at a depth range of 10–100 m. This oceanodromous species often migrates to coastal waters and river mouths in summer and offshore in colder weather and to deep waters during winter.

Maximum length and weight: 103 cm; 12.0 kg

Food and feeding: Its chief food items include shrimps, mollusks, and fishes.

Uses: Commercial fisheries and aquaculture exist for this species. It is marketed fresh or smoked and is highly sought by sport fishermen.

Pharmaceutical and nutraceutical compounds and activities

Antibacterial and hemolytic activities: The mucus of this fish showed antibacterial activity against *V. fluvialis*, *V. alginolyticus*, *Pseudomonas aeruginosa*, and *S. aureus* with a mean diameter of inhibition of 0.80 cm. With *V. parahaemolyticus*, the value was 0.87 cm. The blood serum of this species showed similar activity against *V. alginolyticus* and *V. parahaemolyticus* with 0.80 and 0.87 cm, respectively. The blood serum and mucus also displayed hemolytic effects and the mean diameter values of lysis were found to be 1.35 and 1.00 cm, respectively (Caruso et al., 2014).

Nutritional values: The by-products of this fish with their high n-3 PUFA, especially EPA and DHA; and triglycerides, phospholipids, and total cholesterol serve as potential sources of nutritional supplements (Messina et al., 2013).

2.1.30   THREADFIN BREAMS (ORDER: PERCIFORMES; FAMILY: NEMIPTERIDAE)

***Nemipterus japonicus* (Bloch, 1791)**

Courtesy of. Wikimedia

Common name(s): Japanese thread-fin bream

Global distribution: Tropical; it is native to the Pacific and Indian Oceans; occurs in the Mediterranean

Habitat: This demersal species is found on mud or sand bottoms of coastal waters at a depth range of 5–80 m. It is nonmigratory and is forming schools.

Maximum length and weight: 32.0 cm; 596.0 g

Food and feeding: Its chief food items include crustaceans, mollusks (mainly cephalopods), polychaetes, echinoderms, and other small fishes.

Uses: This species has value of commercial fisheries and is sold fresh, frozen, steamed, dried-salted, and dry-smoked. It is also made into fish balls and fishmeal.

Pharmaceutical and nutraceutical compounds and activities

Antioxidant and antiproliferative effect: The purified peptide fractions derived from the backbone protein hydrolysate of this species showed significant activity against PUFAs peroxidation. Further, these fractions also quenched free radicals (DPPH, hydroxyl, and superoxide). Furthermore, these peptides displayed a significant antiproliferative activity on Hep G2 cell lines (IC$_{50}$, 61.1 g/mL) (Naqash and Nazeer, 2011, 2012).

Naqash and Nazeer (2010) reported that the trypsin protein hydrolysate derived from the muscle of this species showed significant activity against PUFAs peroxidation. Further, the purified peptides of the above hydrolysate demonstrated significant antiproliferative activity on Hep G2 cell lines (Naqash and Nazeer, 2010).

Nutraceutical properties: This species may serve as a functional food due to its rich unsaturated fatty acids such as in linoleic, oleic, and palmitic, and n-3 PUFAs such as EPA and DHA. The values of EPA and DHA were found to be 1.6% and 0.5%, respectively, in the skin and 1.6% and 0.6%, respectively, in the liver (Nazeer et al., 2009).

2.1.31   GOURAMIES (ORDER: PERCIFORMES; FAMILY: OSPHRONEMIDAE)

Trichopodus pectoralis **Regan, 1910**

Source: Image by BEDO (Thailand). https://en.wikipedia.org/wiki/Creative_Commons

Common name(s): Snakeskin gourami

Global distribution: Tropical; Asia: Mekong basin in Laos, Thailand, Cambodia, and Vietnam

Habitat: This benthopelagic, freshwater species lives in the standing-water habitats (4 m depth) with vegetation. It is also known to occur in flooded forests, rivers, and lakes.

Maximum length and weight: 25.0 cm; 500 g

Food and feeding: It feeds mainly on aquatic plants.

Uses: Commercial fisheries and aquaculture exist for this species. Flesh of this species is of good quality and it may be grilled or used for fish soup.

Pharmaceutical and nutraceutical compounds and activities

Nutraceutical properties: The opioid-like neuropeptides, proopiomelano-cortin-derived hormones (i.e., corticotropin, melanotropin, etc.) present in this species have been reported to serve as dietary supplement for anxiety and stress management (Cheung et al., 2015).

2.1.32 BIGEYES (ORDER: PERCIFORMES; FAMILY: PRIACANTHIDAE)

Priacanthus macracanthus **Cuvier, 1829**

Source: Image by Kawahara Keiga, Naturalis Biodiversity Center/Wikimedia Common. Public domain.

Common name(s): Bigeye snapper, red bigeye, spotted bigeye

Global distribution: Subtropical; western Pacific: southern Japan to western Indonesia; Arafura Sea and Australia; Peter the Great Bay

Habitat: This species dwells in inshore and offshore reefs at depths of 20–400 m. It is oceanodromous forming schools in open bottom areas.

Maximum length and weight: 30.0 cm

Food and feeding: Feeds mostly on crustaceans and toleosts.

Uses: This species has commercial fisheries and is often sold fresh.

Pharmaceutical and nutraceutical compounds and activities

Antioxidant activity: The gelatin hydrolysate derived from the pyloric cecae of this species displayed DPPH, and ABTS radical scavenging activities (Phanturat et al., 2010). Nurdiani et al. (2017) reported on the DPPH radical scavenging activity of the peptides from the skin gelatin of this species.

The gelatin hydrolysates from the skin of this species showed DPPH and ABTS radical scavenging activities and ferric reducing antioxidative power. However, hydrolysates derived from gelatin using alcalase and pyloric ceca extract showed the highest ABTS radical scavenging activity (Phantura et al., 2010).

2.1.33   COBIA (ORDER: PERCIFORMES; FAMILY: RACHYCENTRIDAE)

Rachycentron canadum **(Linnaeus, 1766)**

Source: NOAA FishWatch. Public domain.

Common name(s): Cobia

Global distribution: This subtropical fish species is found distributed world-wide in warm marine waters, except for the central and eastern Pacific.

Habitat: This species dwells normally in the coastal and continental shelf waters at a depth range of 0–1200 m. It is also seen in mangrove sloughs, inshore around pilings and buoys, and offshore around drifting and stationary objects. This oceanodromous species may also occasionally visit estuaries and form schools.

Maximum length and weight: 200 cm; 68.0 kg

Food and feeding: It feeds mainly on invertebrates such as crabs, fishes, and squids.

Uses: This species has minor fisheries and it is however known for its commercial aquaculture. As a good food fish, it is sold fresh, smoked, and frozen. It is also a game fish.

Pharmaceutical and nutraceutical compounds and activities

Antioxidant activity: The skin gelatin hydrolysate and enzyme-treated skin gelatin hydrolysate of this species showed DPPH radical scavenging activity (55% and 51–61%, respectively). Further, these hydrolysates showed lipid peroxidation inhibition (58% and 60–71%, respectively) (Yang et al., 2008; Nurdiani et al., 2017).

ACE-inhibitory activity: The head papain hydrolysate (CHPH) and gastrointestinal protease-treated hydrolysate of this species showed significant ACE-inhibitory activity with IC_{50} values of 0.24 and 0.17 mg/mL, respectively. Further, its protease-treated hydrolysate was found to significantly reduce the systolic blood pressure in a dose-dependent manner after oral administration to hypertensive rats at doses of 150, 600, and 1200 mg/kg body weight (Yang et al., 2013).

2.1.34   DRUMS OR CROAKERS (ORDER: PERCIFORMES; FAMILY: SCIAENIDAE)

Argyrosomus japonicus (**Temminck & Schlegel, 1843**) (*=Nibea japonica*)

Common name(s): Giant Croaker, Japanese meagre

Global distribution: Tropical; Indian-West Pacific region

Habitat: This benthopelagic, oceanodromous species usually inhabits the lower niches of river, estuary, reef, and beach area. Adult fish is found mainly near shore beyond the surf zone, occasionally going inshore.

Maximum length and weight: 181 cm; 75 kg

Food and feeding: Its food items include crustaceans, cephalopods, and teleost fishes.

Uses: This species has commercial fisheries with aquaculture and game fish values.

Pharmaceutical and nutraceutical compounds and activities

Antioxidant and wound healing properties: Acid-solubilized and pepsin-solubilized collagens derived from this species showed antioxidant properties. Further, collagen from its swim bladder showed an increased efficacy of wound healing in experimental mice and this was evident by the observed reduction in the levels of interleukin (IL)-1, IL-6, and tumor necrosis factor (TNF) (Chen et al., 2019).

Immunomodulatory activity: The peptide derived from the protein hydro-lysate of this species showed immunomodulatory effects on RAW264.7 cells through the NF-B signaling pathway. This finding suggests that this peptide can be of great use in the production of functional foods for chronic diseases (Zhang et al., 2019).

Cynoscion leiarchus (Cuvier, 1830)

Source: Courtesy of Smithsonian Tropical Research Institute.

Common name(s): Smooth weakfish

Global distribution: Tropical Western Atlantic regions such as Brazil and Nicaragua; and Panama to Santos

Habitat: This estuarine and coastal species is found over mud and sand bottoms at a depth range of 10–50 m.

Maximum length and weight: 90.8 cm; 2.0 kg

Food and feeding: It is a carnivore-feeding species preying on bony fishes and benthic crustaceans, such as crabs and shrimps.

Uses: It has commercial fisheries and its catch is marketed fresh and salted.

Pharmaceutical and nutraceutical compounds and activities

Collagenolytic property: The serine protease derived from the digestive viscera of this species possesses the ability of breaking down type I collagen. This enzyme may also be of great use as a value-added product (Oliveira et al., 2017).

Johnius belangerii (Cuvier, 1830)

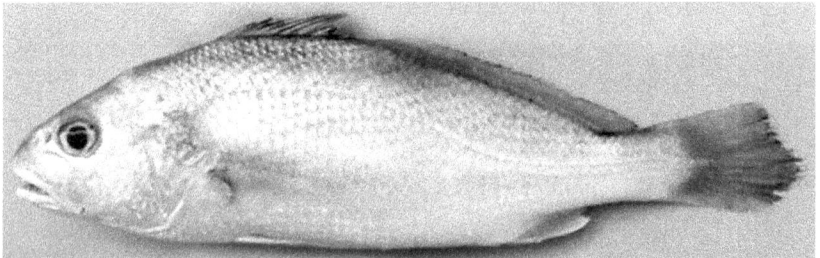

Common name(s): Belanger's croaker

Global distribution: Tropical Indo-West Pacific regions such as India, Sri Lanka, Pakistan, and China

Habitat: This demersal, amphidromous species dwells in the coastal waters and estuaries at a depth range of 0–40 m.

Maximum length: 30.0 cm

Food and feeding: The chief food items of this species include benthic worms, shrimps, crabs, and other fish.

Uses: Exists for this species has minor commercial fisheries and is sold fresh and dried-salted.

Pharmaceutical and nutraceutical compounds and activities

Antioxidant activity: Nurdiani et al. (2017) reported that the skin gelatin peptides of this species with the amino acid sequence as His–Gly–Pro–Leu–Gly–Pro–Leu possessed DPPH radical scavenging activity.

Nutraceutical properties: The fishbone phosphopeptide with 23.6% of phosphorus has shown calcium-binding ability and this could be used as a nutraceutical (Khora, 2013; Wang et al., 2017).

Larimichthys crocea **(Richardson, 1846) (=*Pseudosciaena crocea*)**

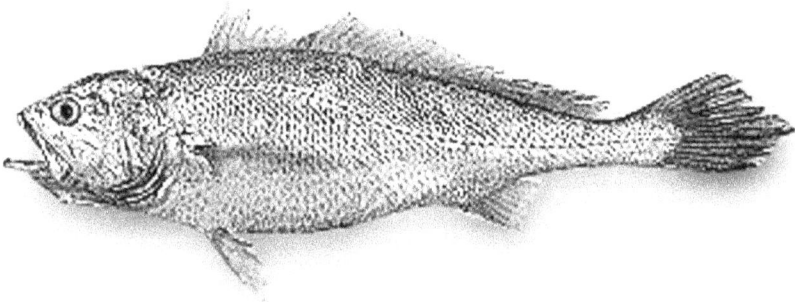

Source: https://www.fao.org/fishery/species/3163/en

Common name(s): Large yellow croaker

Global distribution: Northwest Pacific tropical regions such as East China Sea, and from Vietnam to South Korea and Japan

Habitat: This benthopelagic, oceanodromous species is found in the muddy or muddy–sandy bottoms of coastal waters and estuaries at a depth range of 0–120 m.

Maximum length: 80.0 cm

Food and feeding: It feeds chiefly on crustaceans and fishes.

Uses: This species has high commercial fisheries and is an important aquaculture and aquarium species.

Pharmaceutical and nutraceutical compounds and activities

Antioxidant activity: The protein hydrolysate of the muscle of this species yielded peptides with the amino acid sequence of Tyr–Leu–Met–Ser–Arg (1) and Val–Leu–Tyr–Glu–Glu (2), respectively. Among these peptides, peptide 1 showed significant scavenging activities on DPPH (EC50, 1.35 mg/mL), ABTS (EC50, 0.312 mg/mL) radicals and superoxide (EC50, 0.450 mg/mL); and effective inhibition on lipid peroxidation. On the other hand, peptide 2 demonstrated the strongest hydroxyl radical scavenging activity (EC50, 0.353 mg/mL) (Chi et al., 2015c).

Wang et al. (2013) reported on the isolation of three antioxidant peptides from the scales of this species with the amino acid sequences of GFRGTI-GLVG (P1), GPAGPAG (P2), and GFPSG (P3). These peptides showed scavenging activities on hydroxyl radical with IC_{50} values of 0.293, 0.240, and 0.107 mg/mL, respectively; DPPH radical with 1.271, 0.675, and 0.283 mg/mL, respectively; superoxide radical with 0.463, 0.099, and 0.151 mg/mL, respectively; and ABTS radical with 0.421, 0.309, and 0.210 mg/mL, respectively.

Micropogonias furnieri (Desmarest, 1823)

Common name(s): Whitemouth croaker

Global distribution: Western Atlantic Ocean regions such as Antilles and southern Caribbean coast; and Atlantic coast of South America (Costa Rica to Argentina)

Habitat: It is found usually over muddy and sandy bottoms in coastal waters to about 60 m depths, and estuaries.

Maximum length and weight: 45.0 cm; 55.3 g

Food and feeding: Its chief food items include polychaetes, crustaceans, mollusks, and echinoderms.

Uses: This species has high commercial fisheries. As an important food-fish, it is sold fresh and salted.

Pharmaceutical and nutraceutical compounds and activities

Antioxidant activity: The protein hydrolysates of the muscle of this species exhibited DPPH radical scavenging activity and the percentage values of activity increased in relation to the time of hydrolysis as given in the following table.

Antioxidant activity of muscle of *M. furnieri* hydrolysate hydrolyzed at different durations.

Hydrolysis time (h)	DPPH activity
2	40.4
4	48.0
8	54.1

Source: Lima et al. (2019).

Zavareze et al. (2014) reported that the hydrolysates of the muscle and by-product of this showed lipid peroxidation inhibitory activity and the percentage values were found to be 27% and 31.9%, respectively.

Nutraceutical properties: The soluble fractions of the visceral and muscle hydrolysates of this species showed ideal levels of protein and lipid (86.0% and 0.4%, respectively) and it suggests that the discards of this species may be of great use in the preparation of edible films and functional foods (de Amorim et al., 2016).

Miichthys miiuy (Basilewsky, 1855)

Source: Image by Hsu Chienho. Used with permission.

Common name(s): Miiuy croaker, Chinese drum

Global distribution: Northwest Pacific temperate regions such as Japan to China Sea

Habitat: This demersal, oceanodromous species dwells both in marine and brackish water areas at a depth range of 15–100 m.

Maximum length: 70.0 cm

Food and feeding: The preferred diet of this species is teleost fishes. However, it may also feed crustaceans such as shrimps and crabs.

Uses: This species has commercial fisheries. It is also an important aquaculture species in China.

Pharmaceutical and nutraceutical compounds and activities

Antioxidant properties: The pepsin-soluble collagen (PSC) obtained from this species exhibited significant DPPH, ABTS, hydroxyl radical, and superoxide anion radical scavenging activities. It is suggested that this collagen may be of great use in the manufacture of cosmeceutical products for safeguarding the skin from photoaging or UV damage (Li et al., 2018).

Antibacterial activity: The peptide hepcidin isoform 1 has been isolated from this species (https://www.uniprot.org/uniprot/G0Z5J6). Ke et al. (2015) reported on the antibacterial activity of the synthetic hepcidin on the Gram-positive *Staphylococcus aureus* and Gram-negative *Escherichia coli* and *Aeromonas hydrophila* (Ke et al., 2015).

Cognition enhancement: The swim bladder hydrolysates of this species demonstrated learning and memory enhancement in the experimental mice when the hydrolysate was administered for 28 days. This finding suggests that this species may serve as a prospective food to improve cognition in human (Lu et al., 2010).

Otolithes ruber (Bloch & Schneider, 1801)

Common name(s): Tigertooth croaker, silver teraglin

Global distribution: Indo-West Pacific tropical regions such as East Africa, China Sea, and Queensland

Habitat: This benthopelagic species occurs in coastal and brackish waters at a depth range of 10–40 m. It is amphidromous.

Maximum length and weight: 90.0 cm; 7.0 kg

Food and feeding: While the adult fish feeds mainly on prawns and other teleost fishes, young ones feed on mysiids and Acetes.

Uses: This species has commercial fisheries and it is sold fresh and dried or salted.

Pharmaceutical and nutraceutical compounds and activities

Antioxidant activity: The purified peptide from the muscle protein hydrolysate of this species showed antioxidant activity by scavenging DPPH and hydroxyl radicals at 84.5% and 62.4%, respectively (Khora, 2013; Wang et al., 2017).

Wound healing effect: In experimental Wistar rats, the acid-soluble collagen (ASC) and PSCs derived from the bone of this species acted showed healing effects on excision wounds (Kumar et al., 2012).

2.1.35   MACKEREL, TUNA, AND BONITO (ORDER: PERCIFORMES; FAMILY: SCOMBRIDAE)

Katsuwonus pelamis (Linnaeus, 1758)

Source: NOAA FishWatch. Public domain.

Common name(s): Skipjack tuna

Global distribution: It is a cosmopolitan species with wide distribution in tropical and warm-temperate seas.

Habitat: It is an oceanic and pelagic species living at a depth range of 0–260 m. It is known for its characteristic jumping, feeding, foaming, and so on. It forms schools in in surface waters.

Maximum length and weight: 110 cm; 34.5 kg

Food and feeding: It is a carnivorous species preying on fishes, crustaceans, cephalopods, mollusks, and other fishes. Cannibalism is a common feature in this species.

Uses: This species has highly commercial fisheries exist for this species. It is sold fresh, frozen, dried-salted, and smoked. It has also game fish value.

Pharmaceutical and nutraceutical compounds and activities

ACE-inhibitory activity: The peptides isolated from the bowels of this species displayed ACE-inhibitory activity (IC_{50}, 1 µM) (Pangestuti and Kim, 2017).

Antimicrobial activity: The peptide derived from the acidified skin extract of this fish showed potent activity against Gram-positive bacteria, such as *Bacillus subtilis*, *Micrococcus luteus*, *Staphylococcus aureus*, and *Streptococcus iniae* with minimal effective concentration (MEC) values of 1.2–17.0 g/mL; Gram-negative bacteria, such as *Aeromonas hydrophila*, *Escherichia coli* D31, and *Vibrio parahaemolyticus* (MECs, 3.1–12.0 g/mL); and against *Candida albicans* (MEC, 16.0 g/mL) (Seo et al., 2014).

Rastrelliger kanagurta (Cuvier, 1816)

Common name(s): Indian mackerel

Global distribution: It is widespread in the tropical Indo-West Pacific.

Habitat: This pelagic species is commonly found in coastal bays, harbors, and deep at a depth range of 20–90 m. It is also oceanodromous forming schools.

Maximum length and weight: 42.1 cm; 3.8 kg

Food and feeding: It is a plankton feeder on diatoms, cladocerans, ostracods, and polychaete larvae.

Uses: Highly commercial fisheries exist for this species. This is also a game fish and is rarely used as fish bait.

Pharmaceutical and nutraceutical compounds and activities

Antioxidant activity: The protein hydrolysates derived from the backbones of this species through pepsin and papain digestion showed scavenging of DPPH (46% and 36%, respectively) and superoxide (58.5% and 37.54%, respectively) radicals (Sheriff et al., 2014).

Sarda orientalis (Temminck & Schlegel, 1844)

Source: Image by Hamid Badar Osmany. https://creativecommons.org/licenses/by/3.0/

Common name(s): Oriental bonito; striped bonito

Global distribution: Tropical and subtropical Pacific; eastern Pacific: from Cabo San Lucas, Baja California to the Gulf of Guayaquil; Indo-Pacific; Gulf of California to northern Peru; offshore islands of the tropical eastern Pacific.

Habitat: This pelagic species is found in coastal and ocean waters at a depth range of 1–167. It is also oceanodromous and is schooling with small tunas.

Maximum length and weight: 102 cm; 10.7 kg

Food and feeding: Its main food items include squids, decapod crustaceans, and clupeoid fishes.

Uses: This species has only minor commercial fisheries and is sold fresh, dried-salted, canned, and frozen. It is also a game fish.

Pharmaceutical and nutraceutical compounds and activities

ACE-inhibitory activity: Bagchi et al. (2016) reported on the ACE inhibitory of the peptides from this species.

Clinical observations suggest that a targeted nutritional support program incorporating bonito peptides daily may help to improve blood pressure in borderline and mildly hypertensive patients. The data recorded with three patients are given below.

Observed readings of blood pressure with bonito peptide-fed patients.

	Patient 1		Patient 2		Patient 3	
	0 week	7 weeks	0 week	8 weeks	0 week	12 weeks
Systolic BP (mmHg)	160	138	140	112	133	124
Diastolic BP (mmHg)	89	84	78	70	77	68

Source: Clinical Summary (http://www.metadocs.com/pdf/case_studies/MET1219%20 Bonito%20Peptides%20Case%20Reports.pdf).

Fujita and Yoshikawa (1999) reported that the two peptides (LKPNM and LKP) derived from the thermolysin digest of the dried bonito showed significant antihypertensive (66% and 91%) and weak ACE-inhibitory (0.92% and 7.73%) activities.

Nutraceutical effects: The hydrolysates of this species have been reported to serve as a dietary supplement and functional foods especially for patients with heart (Gevaert et al., 2016).

Scomber australasicus Cuvier, 1832

Source: Courtesy of CSIRO, Australian National Fish Collection

Common name(s): Blue mackerel

Global distribution: Indo-West Pacific subtropical regions such as Red Sea and Persian Gulf; Japan, Australia, and New Zealand; and eastern Pacific: Hawaii and off Mexico

Habitat: This pelagic species is found in coastal and oceanic waters at a depth range of 87–200 m. It is also an oceanodromous species.

Maximum length and weight: 44.0 cm; 1.4 kg

Food and feeding: Its main food items include planktonic copepods and other crustaceans. However, it also feeds on small fish and squids.

Uses: Commercial fisheries exist for this species. It is marketed fresh, dried-salted, smoked, canned, and frozen. This game fish is also used as fish bait.

Pharmaceutical and nutraceutical compounds and activities

Antioxidant activity: The protease N hydrolysate derived from this species showed antioxidant activity and the scavenging effect of this hydrolysate on DPPH radical was 15.4% (Wu et al., 2003).

Scomber japonicus **Temminck & Schlegel, 1782** (*=Pneumatophorus japonicus*)

Source: NOAA FishWatch. Public domain.

Common name(s): Pacific chub mackerel, Spanish mackerel

Global distribution: Temperate and subtropical Indo-Pacific Ocean

Habitat: This epipelagic to mesopelagic species is found in the coastal and continental slope regions at a depth range of 0–300 m. It schools with other fish species such as *Trachurus symmetricus*, *Sarda chiliensis*, and *Sardinops sagax*.

Maximum length and weight: 64.0 cm; 2.9 kg

Food and feeding: Its main food items include copepods, squids, and other fishes.

Uses: This species is known for its commercial and recreational fisheries in California. It is sold fresh, frozen, smoked, salted, and canned. It is also used as bait.

Pharmaceutical and nutraceutical compounds and activities

Antioxidant properties: The visceral hydrolysate of this species displayed in-vitro antioxidant activity on DPPH radicals, hydroxyl radicals, and superoxide anion and the values recorded are given in the following table.

In-vitro antioxidant activities of visceral hydrolysate of *Scomber japonicus*.

Activity	IC_{50} value (mg/mL)
DPPH	0.81
Hydroxyl	0.52
Superoxide	0.46

Source: Wang et al. (2018).

Bashir et al. (2018) and Wang et al. (2017) reported on the antioxidant properties of the muscle protein hydrolysates of this species. Highest DPPH radical scavenging activity (71.69%) was recorded with whole muscle protein hydrolysates treated with Protamex; and the highest ABTS radical scavenging activity (95.39%) was observed in white muscle protein hydrolysates treated with Neutrase. Further, the highest superoxide dismutase (SOD)-like activity (32.84%) was found with white muscle protein hydrolysates treated with Protamex.

Scomber scombrus Linnaeus, 1758

Source: Image by Peter van der Sluijs. https://creativecommons.org/licenses/by-sa/3.0/

Common name(s): Atlantic mackerel

Global distribution: North Atlantic temperate regions including Mediterranean

Habitat: This pelagic species occurs in the coastal and offshore regions at a depth range of 0–1000 m. This oceanodromous species forms large schools near the surface.

Maximum length and weight: 60.0 cm; 3.4 kg

Food and feeding: Its chief food items include zooplankton including fish eggs and larvae, and small fish.

Uses: This species has high commercial fisheries and is sold fresh, frozen, smoked, and canned. It is also a game fish with recreational value.

Pharmaceutical and nutraceutical compounds and activities

Antioxidant, anti-inflammatory, and antihypertensive activities: Hydro-lysates derived from the hydrolysis of the head and skin gelatins with pepsin displayed significant DPPH scavenging activity (80%), and high anti-inflammatory activity (inhibition of SSAO activity by 50%) and antihypertensive activity (inhibition of ACE activity by 75%). The corresponding values for the visceral hydrolysates of this species were 73%, 45.8%, and 60%, respectively. Further, the polar oil fractions derived from the heads after enzymatic hydrolysis with alcalase showed high DPPH radical scavenging activity which was largely due to the presence of carotenoids (Khiari, 2010).

Thunnus albacares **(Bonnaterre, 1788)** (=*Neothunnus macropterus*)

Source: NOAA FishWatch. Public domain.

Common name(s): Yellowfin tuna

Global distribution: Distributed in all tropical and subtropical seas except Mediterranean Sea

Habitat: This epipelagic, oceanic fish is found at a depth range of 1–250 m. It is a strong schooler and is associated with porpoise. Occasionally enters brackish water.

Maximum length and weight: 239 cm; 200 kg

Food and feeding: Its main food items include crustaceans, squids, and fishes.

Uses: This species has high commercial fisheries and is sold fresh, frozen, canned, or smoked. It is a valued fish in sashimi, a Japanese delicacy. It has recreational value as a game fish.

Pharmaceutical and nutraceutical compounds and activities

Antihypertensive (ACE inhibitory) activity: The peptides derived from this species showed ACE-inhibitory activity with an IC_{50} value of 2 µM (Pangestuti and Kim, 2017).

Antibacterial activity: A peptide of 3.4 kDa was purified from an acidified skin extract of this species showed potent antibacterial activity against Gram-positive bacteria, such as *Bacillus subtilis*, *Micrococcus luteus*, and *Streptococcus iniae* (MECs, 1.2–17.0 g/mL), and Gram-negative bacteria, such as *Aeromonas hydrophila*, *Escherichia coli* D31, and *Vibrio parahaemolyticus* (MECs, 3.1–12.0 g/mL) (Seo et al., 2012).

Thunnus obesus (Lowe, 1839)

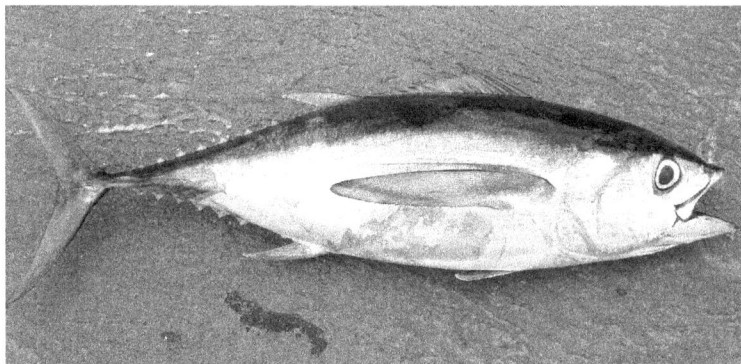

Source: Image by Allen Shimada, NOAA NMFS OST. Public domain.

Common name(s): Bigeye tuna

Global distribution: Tropical and subtropical waters of the Atlantic, Indian, and Pacific Oceans; absent in Mediterranean

Habitat: It is a pelagic–oceanic species occurring at a depth range of 0–1500 m. It is also a highly migratory, oceanodromous species.

Maximum length and weight: 250 cm; 210 kg

Food and feeding: Its main food items include crustaceans, cephalopods, and fishes.

Uses: This species has highly commercial fisheries with game fish value. It is mainly sold canned or frozen. Its meat is highly prized for sashimi in Japan.

Pharmaceutical and nutraceutical compounds and activities

Antioxidant activity: A peptide with the amino acid sequence of H–Leu–Asn–Leu–Pro–Thr–Ala–Val–Tyr–Met–Val–Thr–OH derived from the peptic hydrolysate of the dark muscle of this species. This peptide scavenged effectively four different free radicals, namely, DPPH, hydroxyl, superoxide, and alkyl radical. Further, this peptide also inhibited PUFA peroxidation (Je et al., 2008; Nurdiani et al., 2017).

ACE-inhibitory activity: The pepsin-derived hydrolysate of the dark muscle of this species yielded a peptide which showed significant ACE-I inhibitory activity with an IC_{50} value of 21.6 µM. This peptide when administered at a dose of 10 mg/kg of body weight exerted significant antihypertensive effect in hypertensive rats (Qian et al., 2007).

Thunnus thynnus (Linnaeus, 1758)

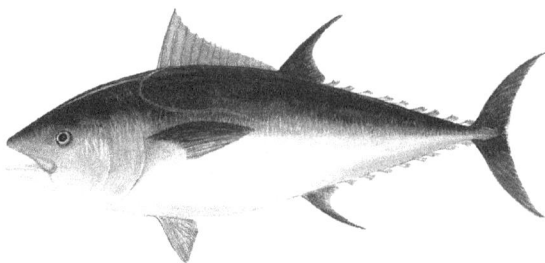

Source: National Oceanic & Atmospheric Administration (NOAA). Public domain.

Common name(s): Atlantic bluefin tuna

Global distribution: Throughout the subtropical and temperate Atlantic and Pacific Oceans; western Atlantic: from Canada to northern Brazil; eastern Atlantic: from Norway to the Canary Islands; western Pacific: from Japan to the Philippines; eastern Pacific: from Alaska to Baja California, Mexico

Habitat: It occupies coastal and pelagic waters at depths from surface to 1000 m. It is a schooler and is found associated with yellowfin, bigeye, skipjack, and albacore.

Maximum length and weight: 458 cm; 684 kg

Food and feeding: It is predator and is preying on small schooling fishes, such as anchovies, sauries, and hakes and on red crabs and squids.

Uses: This species has commercial fisheries and is mainly sold canned; and as fresh fish for sashimi in Japan where it is commercially cultured. It is also a game fish with recreational value.

Pharmaceutical and nutraceutical compounds and activities

ACE-inhibitory activity (antihypertensive activity): The pepsin hydrolysates from the frame of this species exhibited very significant ACE-inhibitory activity (88.2%). A 21-amino acid peptide derived from the pepsin hydrolysate when orally administered to experimental rats with hypertension, resulted in the reduction of systolic blood pressure by 21 mmHg at 6 h (Lee et al., 2010).

Antioxidant and ACE-inhibitory activity: The alcalase hydrolysates from the gelatin of the abdominal skin of this species showed significant antioxidant properties, including metal ion-chelating activity (86.8%), superoxide anion scavenging (39.7%), hydroxyl radical scavenging (37%), and nitrite scavenging (44%) (Han et al., 2015).

Hsu (2010) reported that the peptides isolated from the protease and orientase hydrolysates of the tuna muscle showed DPPH scavenging capacity with the values of 79.6% and 85.2%, respectively.

Yang et al. (2011) reported that the alcalase hydrolysates derived from tuna head showed significant antioxidant effects with IC_{50} values for DPPH, SOD, and hydroxyl radicals as 1.34, 1.2, and 2.84 mg/mL, respectively.

Thunnus tonggol (Bleeker, 1851)

Source: Image by Robbie Cada. Public domain.

Common name(s): Indian Ocean longtail tuna, tongol tuna

Global distribution: Tropical and subtropical; western Indian Ocean to the western Pacific Ocean

Habitat: This epipelagic species is commons in shallow waters of less than 200 m depth. It is however absent in waters with low salinity and high turbidity. It is a schooling fish.

Maximum length and weight: 145 cm; 35.9 kg

Food and feeding: It is an opportunistic feeder and its diet includes crustaceans (particularly stomatopod larvae and prawns), cephalopods, and fishes.

Uses: This species has high commercial fisheries. It is sold mainly fresh, dried-salted, smoked, canned, and frozen. It is also a highly prized game fish.

Pharmaceutical and nutraceutical compounds and activities

Antiproliferative activity: Two different peptides derived from the protein enzymatic hydrolysates of the dark muscle of this species were found to possess antiproliferative activity against MCF7 cells with IC$_{50}$ values of 8.1 and 8.8 M, respectively (Pangestuti and Kim, 2017; Wang et al., 2017).

Nutraceutical properties: The oil from the head of this fish may be of great use as a functional food owing to its monounsaturated (24.7%) and polyunsaturated (26.8%) fatty acids (Ferdosh et al., 2013).

Thunnus sp.

Pharmaceutical and nutraceutical compounds and activities

Antiproliferative activity: The two peptides isolated from the dark muscle hydrolysates of this fish (with the amino acid sequence of Leu–Pro–His–Val–Leu–Thr–Pro–Glu–Ala–Gly–Ala–Thr (1206 Da) and Pro–Thr–Ala–Glu–Gly–Gly–Val–Tyr–Met–Val–Thr (1124 Da)) showed antiproliferative activity against MCF-7 cells (IC_{50}, 8.1 and 8.8 µM, respectively) (Hsu et al., 2011).

2.1.36 GROUPERS AND ROCKCODS (ORDER: PERCIFORMES; FAMILY: SERRANIDAE)

Epinephelus coioides (Hamilton, 1822)

Source: Image by Sahat Ratmuangkhwang. https://creativecommons.org/licenses/by/3.0/

Common name(s): Orange-spotted grouper; estuary cod

Global distribution: Subtropical; from western Indian Ocean to the western Pacific

Habitat: Inhabits coastal reefs at a depth range of 1–100 m. It is eury-thermal and euryhaline.

Maximum length and weight: 120 cm; 32 kg

Food and feeding: It is feeding on small fish, shrimp, and crabs.

Uses: This species has commercial fisheries and aquaculture. It is a highly valued fish for its flesh.

Pharmaceutical and nutraceutical compounds and activities

Chee et al. (2019) reported on the identification of a truncated peptide with 20 amino acids sequence (EGFIFHIIKGLFHAGKMIHG), namely, epine-cidin-1 (Epi-1) from this species and the bioactivities of this compound are given below.

Antibacterial activity: Epinecidin-1 showed significant activity against Gram-negative bacteria such as *Morganella morganii*, *Escherichia coli*, *Vibrio alginolyticus*, *Vibrio parahaemolyticus*, *Pasturella multocida*, *Aeromanas hydrophila*, *Aeromonas sobrio*, and *Flavobacterium menin-gosepticum* with an MBC of less than 2 µg/mL. On the other hand, this compound showed weak activity against other Gram-negative bacteria, namely, *Pseudomonas fluorescens* and *Vibrio vulnificus*, with MBC values of 4.2 and 67.0 µg/mL, respectively.

Antifungal activity: Epi-1 showed fungicidal activity against *Candida albicans*, *Microsporosis canis*, *Trichophytonsis mentagrophytes*, and *Cylindrocarpon* sp. with MIC values of 25, 16.76, 33.52, and 33.52 µg/mL, respectively.

Antiviral activity: Epi-1 exhibited antiviral activity against foot-mouth disease virus at a high concentration of 125 µg/mL. In-vitro studies showed that Epi-1 acted against Japanese encephalitis virus with 40% and 50% at concentrations of 0.5 and 1 µg/mL, respectively.

Anticancer activity: At a concentration of 2.5 µg/mL, the Epi-1 has been reported to show anticancer activity by suppressing the proliferation of U937 cells through apoptosis. Further, this compound inhibited about 90% of the growth of A549, HeLa, and HT1080 cell lines.

Antiparasitic activity: At a concentration of 25 µg/mL, the Epi-1 showed lethal effects against the parasitic protozoan, *Trichomonas vaginalis* strains ATCC 30001, ATCC 50143, and T1.

Immunomodulatory effects: In experimental fishes, the Epi-1 exhibited immunomodulatory activities to protect against bacterial and viral infections. In these fishes, dietary intake of recombinant epinecidin-1 regulated immune-related genes and conferred disease resistance. Further, in zebrafish, this compound gamma-modulated the expressions of immunoresponsive genes like IL-10, IL-1, TNF-, and IFN-γ. Epi-1 also inhibited cytokine production.

Wound healing property: Epi-1 has been reported to heal methicillin-resistant *Staphylococcus aureus* (MRSA)-infected heat burn injuries. This compound is also involved in the repair of neurological injury by inducing the production of glial fibrillary acidic protein.

Epinephelus lanceolatus (Bloch, 1790)

Source: Image by The Cosmonaut. https://creativecommons.org/licenses/by-sa/2.5/ca/deed.en

Common name(s): Giant grouper

Global distribution: Tropical; throughout Indo-Pacific; western Pacific: Japan and Australia; and Oceanic Islands

Habitat: It is benthopelagic and reef-associated and is found in shallow coastal waters and estuaries at a depth range of 1–200 m.

Maximum length and weight: 270 cm TL; 400 kg

Food and feeding: Its main food items include spiny lobsters and fishes such as sharks and batoids; juvenile sea turtles and crustaceans.

Uses: This species has only subsistence fisheries. However, commercial aquaculture exists in Taiwan PC. It is also recreational gamefish. Juveniles possess aquarium value and are sold in ornamental fish trade.

Pharmaceutical and nutraceutical compounds and activities

Anticancer activity: The ultrafiltrated roe hydrolysates of this species possess anticancer activity against the oral cancer cells, Ca9. At 24-h treatment, an IC_{50} value of 0.85 µg/mL was registered for these cells (Yang et al., 2016).

Epinephelus tauvina (Forsskål, 1775)

Source: Image by Hectonichus. https://creativecommons.org/licenses/by-sa/4.0

Common name(s): Greasy grouper

Global distribution: Subtropical; regions of Indo-Pacific; Red Sea and East Africa; islands off North America, South China Sea, Taiwan, and Australia.

Habitat: This species dwells in clear-water areas of coral reefs at a depth range of 1–300 m; and juveniles are normally seen in tidepools, mangrove estuaries, and reef flats. Adults are oceanodromous.

Maximum length and weight: 100.0 cm

Food and feeding: Adults of this species feed exclusively on fishes (holocentrids, mullid, and pomacentrid), and rarely crustaceans

Uses: Minor commercial fisheries and commercial aquaculture exist for this species. As a game fish, it has recreational value.

Pharmaceutical and nutraceutical compounds and activities

Antioxidant activity: The roe protein concentrates prepared from this fish displayed antioxidant activity and the percentage values by DPPH method were 29.4, 34.6, 37.6, 42.8, and 50.7 with 2, 4, 6, 8, and 10 mg concentrations, respectively (Rao, 2014).

2.1.37   SEA BREAMS AND PORGIES (ORDER: PERCIFORMES; FAMILY: SPARIDAE)

Boops boops (Linnaeus, 1758)

Source: Image by riblje-oko.hr. https://creativecommons.org/licenses/by-sa/3.0/

Common name(s): Bogue

Global distribution: Subtropical; Atlantic: Europe, Africa, Mediterranean and Black Seas; Azores and Canary Islands; Norway to Angola

Habitat: This demersal species is found on various bottoms such as sand, mud, rocks, and seaweeds of shelf or coastal areas at a depth range 0–350

m. It is oceanodromous and gregarious, ascending to the surface waters at night.

Maximum length and weight: 40.0 cm; 455.0 g

Food and feeding: It is an omnivorous species feeding mainly on crustaceans and zooplankton.

Uses: This is a game fish with high commercial fisheries value. It is sold fresh, frozen, dried-salted, or smoked. It is also used for fishmeal and oil production and as bait in tuna fisheries.

Pharmaceutical and nutraceutical compounds and activities

Antihyperlipidemic and hepatoprotective effects: Lassoued et al. (2014) reported that the administration of the undigested and digested proteins of this fish to high-cholesterol diet (HCD)-fed rats resulted the reduction of hypercholesterolemia and oxidative stress (Lassoued et al., 2014).

Nutraceutical effects: High levels of amino acids, such as glycine, threonine, arginine, glutamine, histidine, serine, and leucine, have been recorded in the protein hydrolysates of this species and it suggests that his fish could be considered as functional foods for people with malnutrition. The percentage values of the said amino acids recorded in the hydrolysate samples are given below.

Amino acid contents (%) of protein and protein hydrolysate of *B. boops.*

Amino acid	Protein hydrolysate
Glycine	15.8
Threonine	11.0
Arginine	7.6
Glutamine	7.3
Histidine	7.0
Serine	6.8
Leucine	6.5

Source: Lassoued et al. (2014).

Pagellus bogaraveo (Brünnich, 1768)

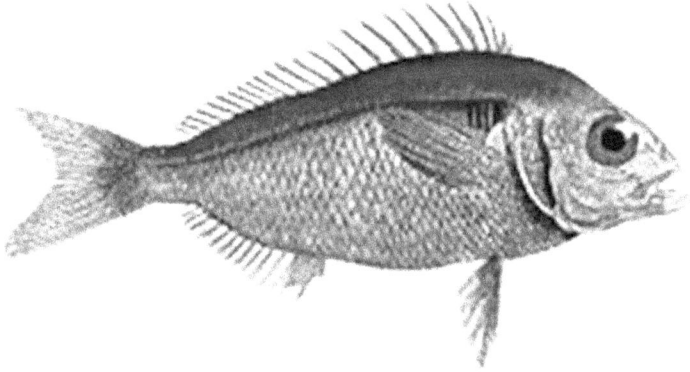

Source: From Georges Léopole Chrétien Frédéric Dagobert, Baron Cuvier, Animal Kingdom, edited by Edward Griffith and printed for George B. Whittaker, London (1827-35). Public domain.

Common name(s): Blackspot seabream

Global distribution: Eastern Atlantic temperate regions such as Norway, Mediterranean, and Iceland

Habitat: This benthopelagic species is found in the inshore waters rocky, sandy, and mud bottoms at a depth range of 150–700 m. While young are seen near the coast, adults are common on the continental slope especially over muddy bottoms.

Maximum length and weight: 70.0 cm; 4.0 kg

Food and feeding: It is an omnivore, feeding mainly on worms, crustaceans, mollusks, and small fish.

Uses: This species has commercial fisheries and it is a game fish with recreational value.

Pharmaceutical and nutraceutical compounds and activities

Antibacterial activity: The mucus of this fish showed antibacterial activity against *V. parahaemolyticus* with a mean diameter of inhibition of 0.80 cm. Its blood serum showed similar activity against *V. anguillarum*, *V. alginolyticus*, and *V. parahaemolyticus* with values of mean diameter of inhibition of 0.80, 1.50, and 1.00 cm, respectively (Caruso et al., 2014).

Hemolytic activity: The mucus of this species displayed hemolytic activity with a mean diameter of lysis of 0.80 cm (Caruso et al., 2014).

***Pagrus major* (Temminck & Schlegel, 1843)** (=*Chrysophrys major*)

Common name(s): Red seabream, Japanese seabream

Global distribution: Subtropical; northwest Pacific: South China Sea; Japan

Habitat: This reef-associated, demersal species lives rough grounds and softer bottoms at a depth range of 10–200 m. It is oceanodromous.

Maximum length and weight: 100.0 cm; 9.7 kg

Food and feeding: Its food items include benthic invertebrates such as echinoderms, worms, mollusks and crustaceans, and occasionally fishes.

Uses: This species has high commercial fisheries and is sold fresh and frozen. It is a popular food fish in Japan. It is also in in commercial aquaculture and in aquarium trade.

Pharmaceutical and nutraceutical compounds and activities

Antibacterial activity: Three isoforms of peptide, namely, Chrysophsin-1 of the pyloric ceca, gills, intestine and stomach; Chrysophsin-2 of gills and stomach; and chrysophsin-3 of gills isolated from this fish species showed antibacterial activity against Gram-negative and Gram-positive bacteria (Saitoh et al., 2019).

Antibacterial and cytotoxic activities: Wang et al. (2012) reported that the cationic peptide Chrysophsin-1 isolated from this species showed broad-spectrum bactericidal activity against both Gram-positive and Gram-negative bacteria and had a significantly lethal effect against *Streptococcus mutans* biofilms which are the causative factors for dental caries and pulpal diseases. Further, this peptide showed cyto-toxic activity against human gingival fibroblasts (HGFs). Toxicity studies, however, showed that chrysophsin-1 has only very weak activity against HGFs.

2.1.38 BARRACUDAS (ORDER: PER CIFORMES; FAMILY: SPHYRAENIDAE)

Sphyraena barracuda (Edwards, 1771)

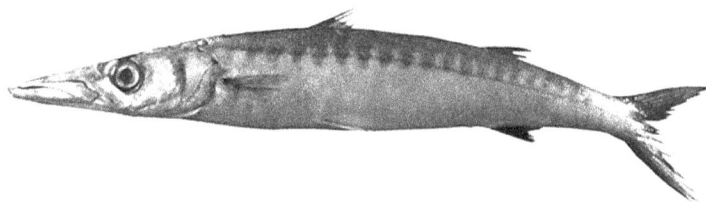

Common name(s): Great barracuda; seela fish

Global distribution: Tropical and subtropical; Indo-Pacific, and Atlantic oceans; from Massachusetts (the United States) to Brazil; Gulf of Mexico and Caribbean Sea; Red Sea

Habitat: This diurnal species lives in clear waters from mangrove areas to deep reef with depth range of 1–100 m. It is solitary or in small aggregations.

Maximum length and weight: 200 cm; 50 kg

Food and feeding: It is a carnivorous fish feeding on fishes, cephalopods, and shrimps.

Uses: This species has only minor commercial fisheries. It is also a game fish with aquarium value.

Pharmaceutical and nutraceutical compounds and activities

Antioxidant activity: Peptides from the backbone hydrolysates of this species showed DPPH and hydroxyl radical scavenging by 61% and 58.7%, respectively (Nurdiani et al., 2017; Nazeer et al., 2011).

2.1.39   SCABBARDFISHES, CUTLASS FISHES (ORDER: PERCIFORMES; FAMILY: TRICHIURIDAE)

Aphanopus carbo Lowe, 1839

Source: Image by R. Mintern. Public domain.

Common name(s): Black scabbardfish

Global distribution: North Atlantic: from the strait of Denmark to western Sahara

Habitat: This bathypelagic and oceanodromous species dwells mainly in deep seas at a depth range of 200–2300 m

Maximum length: 151 cm

Food and feeding: It is a nocturnal feeder migrating to midwater areas to feed on crustaceans, cephalopods, and fishes.

Uses: Highly commercial fisheries exist for this species.

Pharmaceutical and nutraceutical compounds and activities

Antioxidant activity: The protein hydrolysates from the by-products of this species showed DPPH and hydroxyl radical scavenging activity and reducing power. While the hydroxyl scavenging activity decreased with degree of hydrolysis, the other two activities increased with increasing degree of hydrolysis (Batista et al., 2010).

Lepturacanthus savala (Cuvier, 1829)

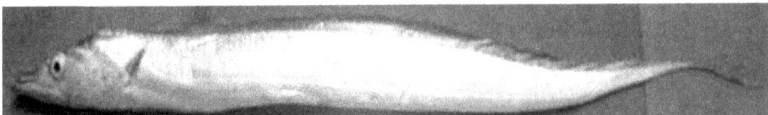

Common name(s): Savalai hairtail

Global distribution: Indo-West Pacific tropical regions such as India, Sri Lanka, the Philippines, Thailand, Malaysia, New Guinea, China, Singapore, Indonesia, and Australia

Habitat: This benthopelagic species dwells in both coastal and brackish water areas at a depth range of 0–100 m. It often comes near the surface at night.

Maximum length and weight: 100.0 cm

Food and feeding: It feeds on small fishes, such as *Setippina, Anchoviella, Harpodon*, and *Trichiurus*, and on prawns.

Uses: Commercial fisheries exist for this species.

Pharmaceutical and nutraceutical compounds and activities

Antioxidant activity: Peptides obtained from the backbone hydrolysates showed DPPH and hydroxyl radical scavenging activity with the values of 60% and 55.6%, respectively (Nurdiani et al., 2017; Nazeer et al., 2011).

Nazeer and Deeptha (2013) also reported that the trypsin hydrolysate of the skin of this species possessed significant DPPH scavenging activity (60%).

Trichiurus lepturus Linnaeus, 1758

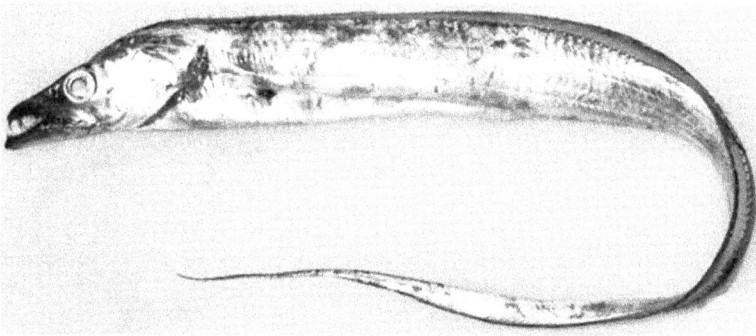

Source: Image byJeff Williams /si.edu. Public domain.

Common name(s): Largehead hairtail

Global distribution: Throughout tropical and temperate seas of the world

Habitat: This benthopelagic species occurs over muddy bottoms from the continental shelf to 350 m depth. It has also been observed occasionally in estuaries.

Maximum length and weight: 234 cm; 5 kg

Food and feeding: Adult fishes are piscivores feeding on anchovies, carangoids, sardines, sphyraenids, sciaenids, and so on, and rarely on crustaceans and molluscs. On the other hand, young feed mostly on euphausiids, planktonic crustaceans, and small fishes.

Uses: Highly commercial fisheries exist for this species has high commercial fisheries and is sold salted, dried, and frozen. It has excellent taste when fried or grilled and used also for sashimi when fresh.

Pharmaceutical and nutraceutical compounds and activities

ACE inhibitory activity: The two peptides of this species derived by trypsin hydrolysis had the amino acid sequences of Ala–Asn–Ser–Glu–Val–Ala–Gln–Trp–Arg (ANSEVAQWR and Glu–Ala–Leu–Val–Ser–Gln–Leu–Thr–Arg (EALVSQLTR), respectively. And both the peptides showed ACE inhibitory activity with IC_{50} values of 89.6 and 91.5 μM, respectively (Fu et al., 2019).

2.1.40　EELPOUT (ORDER: PERCIFORMES; FAMILY: ZOARCIDAE)

Lycodes diapterus Gilbert, 1892

Common name(s): Black eelpout

Global distribution: North Pacific

Habitat: This species is bathydemersal occurring on muddy bottoms at a depth range of 146–844 m

Maximum length: 33.0 cm

Food and feeding: Not reported

Uses: This species has only subsistence fisheries. Its flesh is not normally preferred due to its firm nature. It is therefore used as bait.

Pharmaceutical and nutraceutical compounds and activities

Antioxidative effects: The pepsin hydrolysate prepared from the muscle of this species showed significant antioxidant activity. Further, the peptide purified from this hydrolysate displayed DPPH radical scavenging activity with an EC50 value of was 688.77 µM. Furthermore, this peptide exhibited protective effects against DNA damage induced by oxidation in mouse macrophages (RAW 264.7 cells) (Lee and Byun, 2019).

2.1.41   TONGUE SOLES (ORDER: PLEURONECTIFORMES; FAMILY: CYNOGLOSSIDAE)

Cynoglossus semilaevis **Günther, 1873**

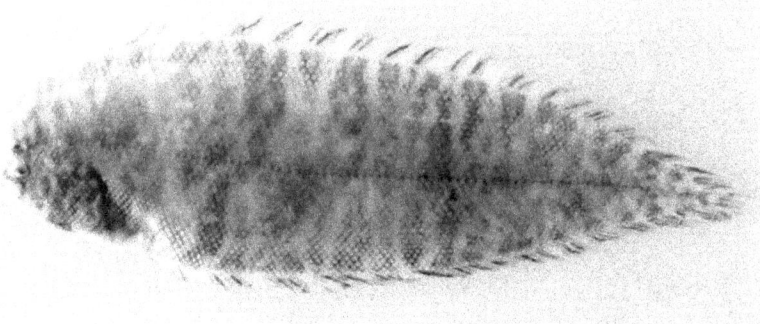

Source: Image by Bedo (Thailand). https://creativecommons.org/licenses/by-sa/4.0/

Common name(s): Tongue sole; half-smooth tongue sole

Global distribution: Subtropical Asian regions such as Yellow Sea and East China Sea

Pharmaceuticals and Nutraceuticals from Fish and Fish Wastes

Habitat: This demersal species inhabits both marine and brackish water areas.

Maximum length: 61.1 cm

Food and feeding: It is a carnivore preying mainly on crustaceans and fish.

Uses: In China, this species has commercial fisheries and aquaculture value.

Pharmaceutical and nutraceutical compounds and activities

Antibacterial and antiviral activity: The peptide, NKLP27, derived from the NK-lysin of this species showed activity against Gram-negative and Gram-positive bacteria, and viral pathogens (Zhang et al., 2014).

2.1.42   FLATFISHES, FLOUNDERS (ORDER: PLEURONECTIFORMES; FAMILY: PARALICHTHYIDAE)

Paralichthys olivaceus (Temminck & Schlegel, 1846)

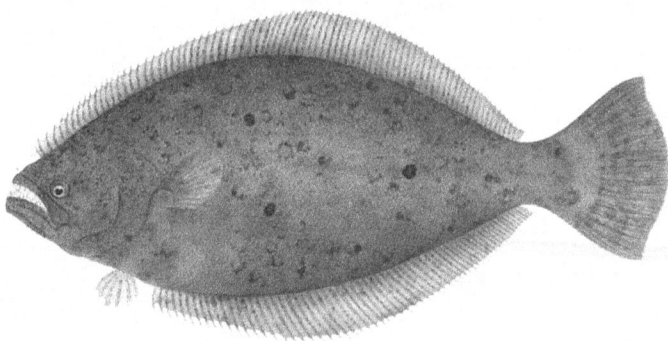

Source: Naturalis Biodiversity Center/Wikimedia Commons. Public domain.

Common name(s): Bastard halibut

Global distribution: Subtropical; western Pacific: from Japan to South China Sea

Habitat: This benthic species occurs mainly on sandy bottoms at a depth range of 10–200 m.

Maximum length and weight: 103 cm; 9.1 kg

Food and feeding: Adults and juveniles of this species feed on other fishes and decapods crustaceans, respectively.

Uses: It is a most highly prized fish species with commercial fisheries and aquaculture values.

Pharmaceutical and nutraceutical compounds and activities

Antioxidant activity: The α-chymotrypsin hydrolysate of this species showed very significant antioxidant activity. Further, the two peptides purified from this hydrolysate showed amino acid sequences of Val–Cys–Ser–Val and Cys–Ala–Ala–Pro, respectively, and had DPPH scavenging activity with IC_{50} values of 111.3 and 26.9 µM, respectively. These peptides also had high cytoprotective activity against 2,2-azobis-(2-amidino-propane) dihydrochloride (Ko et al., 2013).

2.1.43 FLOUNDERS AND HALIBUTS (ORDER: PLEURONECTIFORMES; FAMILY: PLEURONECTIDAE)

Hippoglossus hippoglossus **(Linnaeus, 1758)**

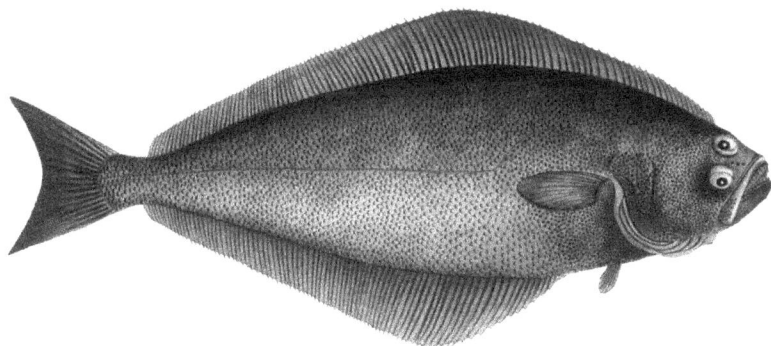

Source: Image by Marcus Elieser Bloch. https://creativecommons.org/licenses/by-sa/4.0/

Common name(s): Atlantic halibut

Global distribution: Temperate and arctic; Northern Atlantic: from Greenland to Iceland; Barents Sea, Bay of Biscay and Virginia, USA

Habitat: This demersal, marine species lives at a depth range of 50–2000 m. Adults are oceanodromous and are rarely pelagic.

Maximum length and weight: 470 cm; 320.0 kg

Food and feeding: Its food items include fishes such as herring, capelin, cod, haddock, sand eels, and pogge); and crustaceans, cephalopods, and other benthic animals.

Uses: Commercial fisheries exist for this species. It is also an aquarium and game fish.

Pharmaceutical and nutraceutical compounds and activities

Antibacterial activity: The peptide, hipposin, isolated from the skin mucus of this species exhibited significant activity against Gram-positive and Gram-negative bacteria and activity at concentrations less than 0.3 µM (Birkemo et al., 2003).

Hippoglossus stenolepis Schmidt, 1904

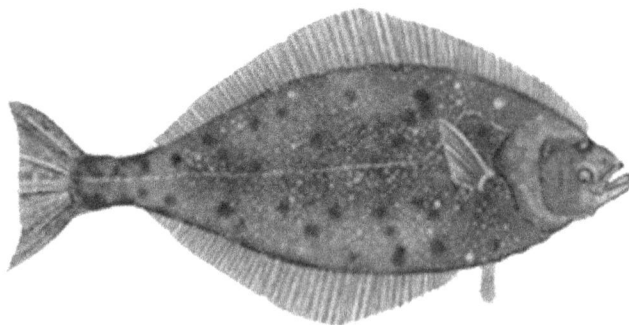

Common name(s): Pacific halibut

Global distribution: Temperate; North Pacific: from Bering Sea to Okhotsk Sea; from Alaska to California; Sea of Japan

Habitat: This benthic and migratory species dwells on different type of bottom types at a depth range of 0–1200 m. While adults prefer deeper water, young and juveniles prefer near shore.

Maximum length and weight: 258 cm; 363.0 kg

Food and feeding: Its food items include crabs, clams, squids, and other fishes.

Uses: This species has commercial fisheries and is sold fresh, dried, salted, smoked, and frozen. It is also an important fish with aquarium and sport values.

Pharmaceutical and nutraceutical compounds and activities

Dipeptidyl-peptidase IV (DPP-IV) inhibitory and antidiabetic activities: The daily administration of this halibut skin gelatin hydrolysate for 30 days improved the glucose tolerance in streptozotocin-induced diabetic rats due to the inhibition of plasma DPP-IV activity, enhanced glucagon-like peptide-1, and insulin secretion (Wang et al., 2015).

Limanda aspera (Pallas, 1814)

Source: Image by Баранчук-Червонный Лев. Public domain.

Common name(s): Yellowfin sole

Global distribution: Temperate; western Pacific: Bering Sea, Korea, Okhotsk Sea, Hokkaido, and Kamchatka; Alaska to Vancouver, Canada

Habitat: This benthic species dwells on sandy bottoms at a depth range of 0–700 m.

Maximum length and weight: 49.0 cm; 1.7 kg

Food and feeding: Adults of this species feed mainly on hydroids, worms, mollusks, and brittle stars

Uses: This species has high commercial fisheries and is sold frozen. It is a game fish too.

Pharmaceutical and nutraceutical compounds and activities

ACE-inhibitory activity: Jung et al. (2006) reported that the peptide (with amino acid sequence, Met–Ile–Phe–Pro–Gly–Ala–Gly–Gly–Pro–Glu–Leu) isolated from the protein hydrolysates of this species showed ACE inhibitory activity with an IC_{50} value of 28.7 µg/mL.

Anticoagulant activity: The protein isolated from this species showed anticoagulant activity by inhibiting the activated coagulation factor XII. At a concentration of 1.0 µM, this protein's activity was recorded at 62% (Rajapakse et al., 2005).

Pleuronectes platessa Linnaeus, 1758

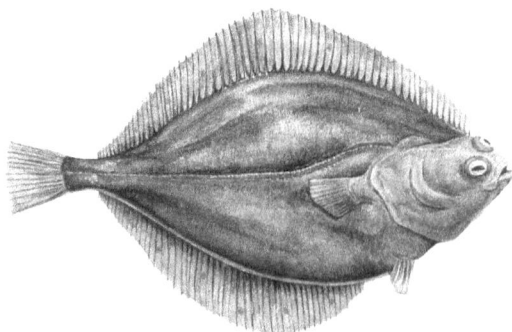

Common name(s): European plaice

Global distribution: Temperate; Northern Sea; Mediterranean Sea

Habitat: This benthic species occurs in sea and estuaries and is rarely entering freshwaters. Its depth range is 0–200 m. Adults live on mixed bottoms and small individuals are usually seen on bathing beaches. It is a resident intertidal species with homing behavior.

Maximum length and weight: 100.0 cm; 7.0 kg

Food and feeding: Its food items include mainly polychaetes and thin-shelled mollusks.

Uses: Highly commercial fisheries exist for this species. It is also a commercially important aquaculture and aquarium species.

Pharmaceutical and nutraceutical compounds and activities

Antiproliferative activity: The protein hydrolysate of this species induced growth inhibition on two human breast cancer cell lines, namely, MDA-MB-231 and MCF-7/6 (Picot et al., 2006).

Antibacterial activity: The protein which is designated as KilC (bacterial killing metalloprotease C) isolated from this species exhibited 74% of activity against *Staphylococcus aureus*, *Escherichia coli*, *Bacillus subtilis*, and *Pseudomonas aeruginosa* efficiently (Tvete and Haugan, 2008).

Pseudopleuronectes americanus **(Walbaum, 1792)** (*=Pleuronectes americanus*)

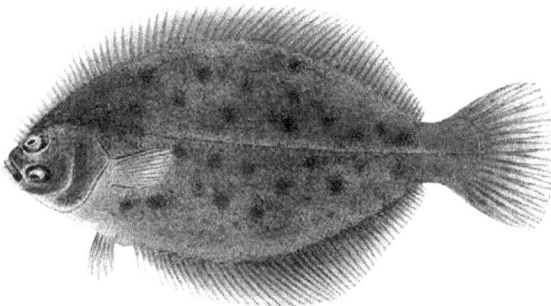

Common name(s): Winter flounder

Global distribution: Western Atlantic regions of temperate such as Canada to Georgia; USA and Labrador

Habitat: This benthic species inhabits soft muddy to fairly hard bottoms at a depth range of 5–143 m. It is oceanodromous.

Maximum length and weight: 64.0 cm; 3.6 kg

Food and feeding: This species feeds mainly on shrimps, amphipods, crabs, snails, and sea urchins.

Uses: This species has commercial fisheries and is presently under experimental aquaculture.

Pharmaceutical and nutraceutical compounds and activities

Antimicrobial activity: The alanine-rich peptide, known as Pa-MAP, derived from this species has been reported to prevent *E. coli* infection and increase mice survival at concentrations of 1 and 5 mg/kg (Teixeira et al., 2013).

Migliolo et al. (2016) reported that the palindromic peptide (Pa-MAP2) obtained from this species displayed activity against Gram-negative bacteria, with an MIC value of 3.2 μM. In-vivo studies showed that Pa-MAP2 increased to 100% the survival rate of mice infected with *Escherichia coli*. These findings indicate that palindromic Pa-MAP2 could be an alternative candidate for use in therapeutics against Gram-negative bacterial infections.

Cole et al. (1997) reported that the purified peptide pleurocidin derived from the skin mucous secretions of this species showed antibacterial activity on certain bacterial species and the MIC values are given below.

	MIC (μg/mL)
S. aureus	17.7
E. coli	2.2
S. typhimurium (I and II)	8.8
B. subtilis	1.1
P. haemolytica	4.4
A. salmonicida	17.7

Source: Cole et al. (1997).

2.1.44 SOLES (ORDER: PLEURONECTIFORMES; FAMILY: SOLEIDAE)

Pardachirus marmoratus (Lacepède, 1802)

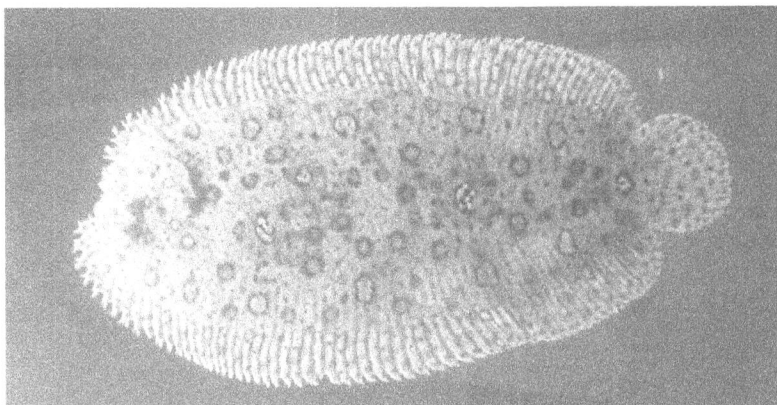

Common name(s): Finless sole

Global distribution: Tropical; Western Indian Ocean: from Red Sea to Durban; and Sri Lanka

Habitat: This species occurs in the sand and mud bottoms of shallow coastal waters and coral reef areas at a depth range of 1–15 m.

Maximum length: 26.0 cm

Food and feeding: It feeds mainly on bottom-living invertebrates.

Uses: Commercial fisheries exist for this species.

Pharmaceutical and nutraceutical compounds and activities

Antibacterial and hemolytic activity: Pardaxin, a 33-amino-acid pore-forming polypeptide toxin isolated from this species, displayed activity against bacteria and human erythrocytes. The values of MIC of pardaxin and its various analogs on the different bacterial species are given in the following table. Pardaxin was found to inhibit the growth of both Gram-positive and Gram-negative bacteria at micromolar concentrations (Oren and Shai, 1996).

Antibacterial activity of pardaxin and its analogs on bacterial species.

Peptide	MIC (μM)						
	Ec	Ac	Pa	St	Bm	Ml	Bs
Paradaxin	13	3	25	40	3	10	5
Pardaxin-[NH(CH$_2$)NH$_2$]$_2$	3	2	8	15	0.8	2	1.2
Lys-des-(23–33)-pardaxin-NH(CH$_2$)NH$_2$]	23	3	12.5	12.5	0.75	1.5	1.2
Des-(23–33)-pardaxin-NH(CH$_2$)NH$_2$]$_2$	8	2.5	50	40	0.8	2	1.2

EC: *E. coli* (D21); Ac: *A. calcoaceticus* (Ac11); Pa: *P. aeruginosa* (ATCC 27853); St: *S. typhirnurium* (LT2); Bm: *B. rnegateriurn* (Bm11); Ml: *M. luteus* (ATCC 9341); Bs: *B. subtilis* (ATCC 6051).

Source: Oren and Shai (1996).

2.1.45 SALMON (ORDER: SALMONIFORMES; FAMILY: SALMONIDAE)

Oncorhynchus gorbuscha (Walbaum, 1792)

Source: A. Hoen and Co. - Scanned from plates in Evermann, Barton Warren; Goldsborough, Edmund Lee (1907) The Fishes of Alaska, Washington, D.C.: Department of Commerce and Labor Bureau of Fisheries. Public domain.

Common name(s): Pink salmon

Global distribution: Subtropical; Pacific: Korean; Hokkaido to northern California; and from Lena River to Mackenzie River in Arctic Sea

Habitat: This benthic species lives in ocean and coastal streams; also in freshwater and brackish water areas at a depth range of 0–250 m.

Maximum length and weight: 76.0 cm; 6.8 kg

Food and feeding: In the sea, young feed on copepods and larvacean tunicates and adults on euphausiids, amphipods, and fishes. Cirripeds, ostracods, tunicates, and dipterous insects may also be the other food items of this species.

Uses: Highly commercial fisheries exist for this species. It is also a commercially important aquaculture species. It is a game fish too.

Pharmaceutical and nutraceutical compounds and activities

ACE-inhibitory activity: A total of 20 active di and tripeptides, including Ile–Val–Phe and Phe–Ile–Ala, have been identified from the muscle of this species. Among them, the Ile–Trp showed significant ACE-inhibitory activity with an IC_{50} value of 1.2 μM (Enari et al., 2008; Pangestuti and Kim, 2017).

Oncorhynchus keta (Walbaum, 1792)

Source: Image by Timothy Knepp / U.S. Fish and Wildlife Service. Public domain.

Common name(s): Chum salmon

Global distribution: Temperate; North Pacific: Alaska to California (USA); British Columbia (Canada), Korea, Japan, and Bering seas; Arctic: Laptev Sea to the Beaufort Sea

Habitat: This benthopelagic, anadromous species lives in marine, freshwater, and brackish water areas; depth range is 0–250 m

Maximum length and weight: 100.0 cm; 15.9 kg

Food and feeding: Its food items include copepods, euphausiids, pteropods, squid, tunicates, and small fishes. Adults do not normally feed in freshwater.

Uses: Highly commercial fisheries exist for this species has commercial fisheries. It also has aquaculture, aquarium, and game fish values.

Pharmaceutical and nutraceutical compounds and activities

ACE-inhibitory activity: The peptide with the amino acid sequence of Gly–Leu–Pro–Leu–Asn–Leu–Pro obtained from the trypsin hydrolysate of the skin of this species showed significant ACE-inhibitory activity with an IC_{50} value of 18.7 µM. Further, in experimental rats, oral administration of synthesized peptide resulted considerable reduction in systolic blood pressure (Lee et al., 2014).

Anti-Alzheimer's and neuroprotective activity: Peptides derived from the skin collagen hydrolysate of this species showed anti-Alzheimer by improving the long-term learning in experimental rats by reducing oxidative damage and acetylcholinesterase activity. Further, this peptide showed neuroprotective activity by increasing neurotrophic factor expression in the brain (Xu et al., 2015).

Wound healing properties: The experimental rats treated with the peptides of this species showed quick wound healing and improved tissue regeneration. Further, this treatment helped in the formation of better organized collagen fiber deposition (Zhang et al., 2011).

Oncorhynchus mykiss (Walbaum, 1792)

Source: Image by Eric Engbretson / Eric / U.S. Fish and Wildlife Service. Public domain.

Common name(s): Rainbow trout

Global distribution: It has subtropical distribution with Armenia, Canada, Mexico, Russia, and USA.

Habitat: This benthopelagic, anadromous species inhabits marine, freshwater, and brackish water areas with a depth range of 0–200 m. It is however common in clear, cold headwaters, creeks, small-to-large rivers, lakes, and intertidal areas.

Maximum length and weight: 122 cm; 25.4 kg

Food and feeding: Its food items include a variety of aquatic and terrestrial invertebrates and small fishes. However, cephalopods and fishes are largely preyed by this species at sea.

Uses: This species has commercial fisheries and is sold fresh, smoked, canned, and frozen. It is also known for its commercial aquaculture and game fish values.

Pharmaceutical and nutraceutical compounds and activities

ACE-inhibitory activity: The hydrolysate of this species obtained with pepsin, trypin, and chymotrypsin showed significant ACE-inhibitory activity with the values of 0.61, 1.09, and 1.51 μg/mL, respectively. Further, the fraction with the amino acid sequence Lys–Val–Asn–Gly–Pro–Ala–Met–Ser–Pro–Asn–Ala–Asn separated from pepsin hydrolysates displayed very significant ACE-inhibitory activity with an IC_{50} value of 0.19 μg/mL (Kim and Byun, 2012).

Antioxidant activity and anticancer effects: The skin protein hydrolysates of this species demonstrated DPPH radical inhibitory power and ferric reducing power. Further, the fractions of these hydrolysates showed cytotoxic properties against HCT-116 cancer cells and inhibited the growth of these cells in vitro (Yaghoubzadeh et al., 2019).

Salmo salar Linnaeus, 1758

Source: Image by Timothy Knepp / U.S. Fish and Wildlife Service. Public domain.

Common name(s): Atlantic salmon

Global distribution: This species occurs in all countries with rivers flowing into the North Atlantic Ocean and the Baltic Sea. In the Western

Atlantic, it is seen abundantly in Quebec, Canada; Connecticut and New York, USA.

Habitat: It inhabits freshwater for about 2 or 3 years in the beginning of its life before entering the sea. It prefers gravelly bottom of rivers during its early life. Smolts enter sea where they live for about 2 years before returning to their freshwater habitats. This amphihaline species has a depth range of 0–210 m.

Maximum length and weight: 150 cm; 46.8 kg

Food and feeding: Adults of this species feed on squids, shrimps, and fish at sea. Juveniles, however, feed mainly on aquatic mollusks, crustaceans, and fish.

Uses: Highly commercial fisheries exist for this species has commercial fisheries and it also has aquaculture and game fish values.

Pharmaceutical and nutraceutical compounds and activities

ACE-inhibitory activity: Peptides of this species obtained with ex-vivo and in-vitro hydrolysis exhibited significant ACE-inhibitory activity with IC_{50} values of 2.16 and 1.04 mg/mL, respectively (Darewicz et al., 2014).

Antioxidant properties: Peptides derived through gastric digest of this species showed significant DPPH radical and 2,2'-azino-bis(3-ethylben-zothiazoline-6-sulfonic acid (ABTS) scavenging activities with the values of 8.9% and 72.7%, respectively, and ferric ion-reducing activity (>80%) (Borawska et al., 2016a).

Falkenberg et al. (2014) reported that the peptides derived from the gills, belly flap muscle, and skin extracts of this species showed ABTS radical scavenging activities and their EC50 (half maximal effective concentration) values are given below:

ABTS radical scavenging activities (EC50 (mg/mL).

Skin	1.2×10^{-2}
Muscle	8.3×10^{-3}
Gills	5.2×10^{-3}

Source: Falkenberg et al. (2014).

Anticoagulation (antithrombin) activity: This species has been reported to exhibit anticoagulation activity owing to its antithrombin content which varies between 1.5 and 6 units/mL (Khora, 2013).

Antiproliferative activity: The peptides derived from the hydrolysate of this species showed antiproliferative activity on two human breast cancer cell line MCF-7/6 and MDA-MB-231 by inducing growth inhibition (Picot et al., 2006).

2.1.46   FLATHEADS (ORDER: SCORPAENIFORMES; FAMILY: PLATYCEPHALIDAE)

Platycephalus fuscus Cuvier, 1829

Source: Image by Richard Ling from NSW, Australia. https://creativecommons.org/licenses/by-sa/2.0/deed.en

Common name(s): Dusky flathead

Global distribution: It is endemic to Australia and it has subtropical western Pacific distribution.

Habitat: This demersal species generally lives in shallow bays and inlets and it can be found in estuaries as far as tidal limits. Its depth range is 0–30 m. Intertidal areas of mud, silt gravel, sand, and seagrass (mainly *Zostera* species) beds are also often inhabited by this species. It may enter freshwater also.

Maximum length and weight: 120 cm; 15.0 kg

Food and feeding: Its common food items are worms, crabs, prawns, octopus, squid, and other small fish.

Uses: This species has commercial fisheries with game fish value.

Pharmaceutical and nutraceutical compounds and activities

Antioxidant and cytotoxic activities: The peptide fractions derived from the byproducts of this species showed the highest DPPH and ABTS radical scavenging activities with the values of 94.0% and 82.9%, respectively. Further, these peptides exhibited significant cytotoxic activities by inhibiting the growth of HT-29 colon cancer cells up to 91.04% (Nurdiani et al., 2016).

2.1.47   SCORPIONFISH (ORDER: SCOR PAENIFORMES; FAMILY: SCORPAENIDAE)

Pterois volitans **(Linnaeus, 1758)**

Source: Image by Alexander Vasenin. https://creativecommons.org/licenses/by-sa/3.0/

Common name(s): Red lionfish

Global distribution: Tropical; Western North Atlantic, Gulf of Mexico, and Caribbean Sea

Habitat: This species favors reef habitats at a depth range of 10–175 m, although it is found near sandy bottoms, mangrove, and seagrass, and lagoons and harbor.

Maximum length: 38.0 cm

Food and feeding: As a nocturnal predator, it preys on shrimps, crabs, and small fishes.

Uses: This species has commercial fisheries and it is popular in ornamental fish trade.

Pharmaceutical and nutraceutical compounds and activities

Antioxidant and antiproliferative activity: The backbone of this species yielded peptic protein hydrolysate which displayed significant free radical scavenging and lipid peroxidation inhibitory activities. The further purified fraction of the above hydrolysate had pronounced antiproliferative activity on Hep G2 cell lines (Naqash and Nazeer, 2011).

Hemolytic activity: The venom of this species has been reported to possess hemolytic activity on the erythrocytes of horse, sheep, cattle, guinea pig, and chicken and the activity was recorded at 23,700 hemolytic units/mg of venom (Ziegman and Alewood, 2015).

Antinociceptive activity: The venom of this species showed antinociceptive activity by increasing the activity of Na$^+$, K$^+$, and ATPase, pain-mediating agents (Ziegman and Alewood, 2015).

Anticancer activity: The peptide derived from this species showed anti-cancer activity in HEp2 and HeLa cells by selectively inducing apoptosis (Ziegman and Alewood, 2015).

Others: The venom of this species is believed to contain acetylcholine, although its role is unknown (Ziegman and Alewood, 2015)

Scorpaena notata Rafinesque, 1810

Source: Image by Kókay Szabolcs. https://creativecommons.org/licenses/by-sa/3.0/deed.en

Common name(s): Small red scorpionfish

Global distribution: Subtropical; Eastern Atlantic

Habitat: This benthic species inhabits rocky littoral areas at depths, 10–700 m

Maximum length: 26.0 cm

Food and feeding: Feeds on crustaceans and small fishes

Uses: This species has commercial fisheries. It is also an important aquarium fish. Its flesh is tasty and is used in making "bouillabaisse."

Pharmaceutical and nutraceutical compounds and activities

ACE-inhibitory and antioxidant activities: The peptides (A–F) derived from the protein hydrolysates of this species showed ACE-inhibitory and DPPH scavenging activities and the values recorded for both activities at a concentration of 10 µg/mL of the peptide fractions are given below (Aissaoui et al., 2015).

Peptide	ACE inhibition (%)	DPPH scavenging activity (%)
A	13.7	11.8
B	3.3	3.3
C	13.6	2.1
D	81.4	18.4
E	63.9	15.3
F	83.3	17.3

Source: Aissaoui et al. (2015).

The protein hydrolysate obtained from the muscle of this fish with the protease of the fungus *Penicillium digitatum* yielded two peptides with the amino acid sequence of Leu–Val–Thr–Gly–Asp–Asp–Lys–Thr–Asn–Leu–Lys and Asp–Thr–Gly–Ser–Asp–Lys–Lys–Gln–Leu, respectively. These peptides with their hydrophilic amino acids showed significant antioxidative and ACE inhibitory activities (Aissaoui et al., 2017a).

Nutraceutical properties: The enzyme protease derived from the intestine of this species has been reported to be of great use in the preparation of food protein hydrolysates (Aissaoui et al., 2017b).

2.1.48 ROCKFISHES (ORDER: SCORPAENIFORMES; FAMILY: SEBASTIDAE)

Sebastes schlegelii Hilgendorf, 1880

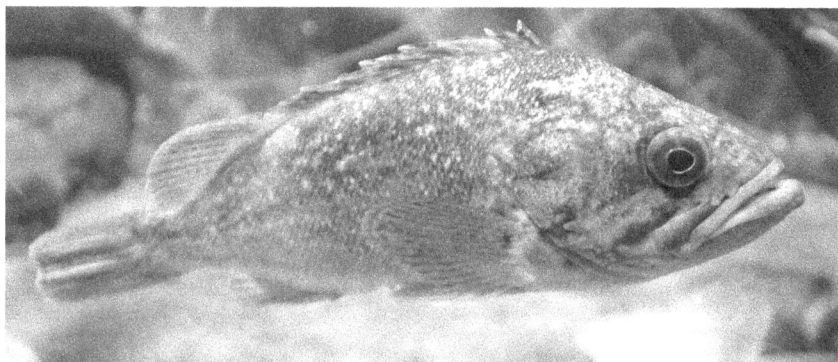

Source: Image by Tomarin. https://creativecommons.org/licenses/by-sa/2.1/jp/deed.en

Common name(s): Black rockfish, Korean rockfish

Global distribution: It has temperate northwest Pacific distribution with countries such as Japan, Korean Peninsula, and China.

Habitat: This benthic species is normally found near shore and on rock bottoms at a depth range of 3–100 m. Its young are often found with drifting seaweeds.

Maximum length and weight: 65.0 cm; 3.1 kg

Food and feeding: Not reported.

Uses: It is a commercial aquaculture species in East Asia and Japan. It is also an aquarium species.

Pharmaceutical and nutraceutical compounds and activities

Antibacterial activity: The glycoprotein, L-amino acid oxidase derived from the skin mucus of this fish showed potent antibacterial activity against *Aeromonas salmonicida*, *Photobacterium damselae* subsp. *piscicida*, and *Vibrio parahaemolyticus* with MIC values of 0.08, 0.2, and 0.6 µg/mL, respectively (Kitani et al., 2007).

2.1.49   STONEFISH, STINGFISH (ORDER: SCORPAENIFORMES; FAMILY: SYNANCEIIDAE)

Choridactylus multibarbus **Richardson, 1848**

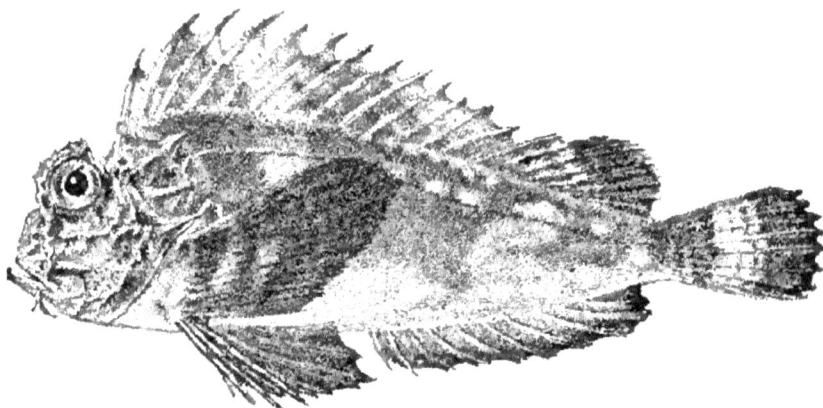

Source: Image by Sir Francis Day - Fauna of British India (Fishes, Volumes 1 and 2). Public domain.

Common name(s): Orangebanded stingfish

Global distribution: It has Indo-West Pacific distribution with Red Sea and Persian Gulf; Thailand, China, the Philippines, India, and Pakistan.

Habitat: This marine, demersal species inhabits sand or mud bottoms at a depth of about 40 m. It is a highly venomous species.

Maximum length: 12.0 cm

Food and feeding: Ghost shrimp is predominantly eaten by this species.

Uses: This species has minor commercial fisheries owing to its venomous nature. It has food value only in a few areas of its geographical distribution.

Pharmaceutical and nutraceutical compounds and activities

Antimicrobial activity: The spine venom of this species showed mild activity against bacterial strains, namely, *Staphylococcus aureus*, *Pseudomonas aeruginosa*, *E. coli*, *Bacillus* sp., *Lactobacillus brevis*, *Vibrio* sp.; and fungal strains, namely, *A. flavus*, *A. niger*, *Candida albicans*, *A. oryzae*, and *A. sojae* (Prithiviraj et al., 2015).

Hemolytic activity: The spine venom of this fish-induced hemolysis in chick, goat, sheep, and human blood samples. The hemolytic titer values in case of chick and goat crude extracts were found to be 5 and 4, respectively; and the values of their specific hemolytic activity were 32 HT/mg of protein and 16 HT/mg of protein, respectively. The corresponding values for sheep and human blood crude extracts were 4 and 3, respectively; and 16 HT/mg of protein and 6 HT/mg of protein, respectively (Prithiviraj et al., 2015).

Cytolytic activity: The spine venom of this species showed potent cytolytic activity against Vero and Mcf 7 cell lines and their viability was adversely affected. Further, this spine venom was found to induce more serious morphological alteration in both the cell lines (Prithiviraj et al., 2015).

Synanceia horrida (Linnaeus, 1766)

Source: George Henry Ford - Day, Francis (1878) The Fishes of India. Volume 2. Public domain.

Common name(s): Estuarine stonefish

Global distribution: It is common in the tropical Indo-West Pacific; from India to Australia; and China.

Habitat: It occurs on sandy, muddy, silty, or sandy bottoms of coastal reefs and estuaries at a depth range of 0–40 m. It is nocturnal and is lying motionless during day time.

Maximum length: 60.0 cm

Food and feeding: It is a nocturnal carnivore feeding on small fishes, crustaceans, and cephalopod mollusks.

Uses: It has no fisheries value owing to its venomosity. But it is often used in public aquariums.

Pharmaceutical and nutraceutical compounds and activities

Antibacterial and hemolytic activity: The aqueous and methanol crude venom (stonustoxin) extracts of this species have been reported to possess very low antibacterial activity. The aqueous and methanol extracts inhibited the growth of Vibrio *cholerae* and *Pseudomonas* sp., respectively. Both the crude extracts exhibited potent hemolytic activity (in chick blood erythrocytes) which was estimated as 8.9 HT/mL for aqueous extract and 12.49 HT/mL for methanol extract (Prithiviraj et al., 2012). Chen et al. (1997) also reported that its stonustoxin induced potent hemolytic activity, that is, lysis of erythrocytes through the formation of pores in the cell membrane.

Cardiovascular activity: The stonustoxin of this species exerted cardiovascular activity with positive effect on blood pressure in anesthetized rabbits and rats (Ziegman and Alewood, 2015).

Synanceia verrucosa Bloch & Schneider, 1801

Common name(s): Reef stonefish

Global distribution: Tropical Pacific and Indian Oceans; from Red Sea to Great Barrier Reef

Habitat: This species dwells in shallow coral bottoms. It may also be seen on and around rocks and plants.

Maximum weight: 2.4 kg

Food and feeding: It feeds mainly on small fish, shrimp, and other crustaceans.

Uses: It has minor commercial fisheries owing to its venomous nature. However, it is eaten in Hong Kong, the Philippines, China, and Japan. It is an important aquarium fish.

Pharmaceutical and nutraceutical compounds and activities

Cytotoxic activity: The crude dorsal venom of the dorsal spines of this species had weak cytotoxic activity on murine P388 leukemia cells (Kato et al., 2016).

Hemolytic activity: The toxins of its venom, namely, verrucotoxin, neover-rucotoxin, and cardioleputin have been reported to exhibit hemolytic activity in experimental rabbits (Ziegman and Alewood, 2015).

2.1.50   CATFISH (ORDER: SILURIFORMES; FAMILY: ARIIDAE)

Arius dussumieri (Valenciennes, 1840)

Common name(s): Blacktip sea catfish

Global distribution: Tropical; Indo-Pacific regions such as Mozambique, India, and Bangladesh; Myanmar to Sumatra; Madagascar to Sri Lanka

Habitat: This demersal species inhabits marine, freshwater, and brackish water areas at a depth range of 20–50 m. Adult fishes enter the lower parts of rivers.

Maximum length and weight: 62.0 cm; 1.4 kg

Food and feeding: Its main food items include small fishes and invertebrates.

Uses: It is marketed fresh and dried-salted, and air bladder is utilized for the preparation of isinglass. In the clarification of beer and wine, the isinglass (collagen) is largely used.

Pharmaceutical and nutraceutical compounds and activities

Hemolytic and hemagglutinating activities: An hemolytic activity from 4 to 32 HU has been observed from the partially purified fractions of this species. Further, the highest hemagglutinating titer of 32 HAU was also observed in this species. The crude mucus extracts of this species also showed a higher amount of edematous activity (154.29%) (Deo et al., 2008).

Nemapteryx caelata (Valenciennes, 1840)

Source: Robert Mintern - Day, Francis (1878) The Fishes of India. Volume 2. Public domain.

Common name(s): Engraved catfish

Global distribution: South and southeast Asia of Indo-Pacific: Sri Lanka, Malaysia, Cambodia, Bangladesh, Pakistan, India, Myanmar, Thailand, and Indonesia

Habitat: This benthic species is common in marine habitats although it enters brackish waters and tidal rivers. As an amphidromous, it ascends into freshwater also.

Maximum length: 45.0 cm

Food and feeding: Its major food items include small fishes and invertebrates.

Uses: This species has commercial fisheries and is sold fresh and dried-salted.

Pharmaceutical and nutraceutical compounds and activities

Antioxidant activities: The hydrolysates derived from the roe of this catfish showed antioxidant activity.

At concentration of 1 mg/mL, the soluble hydrolysates formed after the first stage of hydrolysis and second stage of hydrolysis showed DPPH scavenging activity with the percentage values of 16.8 and 19.0, respectively (Binsi et al., 2016).

Osteogeneiosus militaris (**Linnaeus, 1758**)

Source: Image by Staticd. https://creativecommons.org/licenses/by-sa/3.0/

Common name(s): Soldier catfish

Global distribution: Tropical; Indo-West Pacific: India, Java, Borneo, and Sumatra

Habitat: This benthic species is found in all aquatic habitats at a depth range of 0–32 m. It is potamodromous.

Maximum length and weight: 39.9 cm

Food and feeding: Its main food items include small fishes and invertebrates.

Uses: This species has commercial fisheries.

Pharmaceutical and nutraceutical compounds and activities

Bioactivities: The partially purified fractions of this species showed highest hemolytic activity of 16 HU and hemagglutinating activity of 32 HAU (Deo et al., 2008).

2.1.51   CATFISH (ORDER: SILURIFORMES; FAMILY: BAGRIDAE)

Tachysurus fulvidraco (**Richardson, 1846**) (=*Pelteobagrus fulvidraco*)

Source: Image by A. C. Tatarinov. https://creativecommons.org/licenses/by-sa/4.0/deed.en

Common name(s): Yellow catfish

Global distribution: Temperate; Eastern Asia: Siberia to China, Laos, Korea, and Vietnam

Habitat: This nonmigratory, benthic species is mainly found in lakes and river channels.

Maximum length and weight: 34.5 cm; 300 g

Food and feeding: Its main food items include mollusks, fishes, and insects like trichopterans and chironomids.

Uses: This species has minor commercial fisheries. As an important food-fish, it also has aquaculture value.

Pharmaceutical and nutraceutical compounds and activities

Antimicrobial activity: A peptide, pelteobagrin derived from the skin mucus of this fish displayed antibacterial and antifungal activities and the MIC values obtained for these species different are given below.

Pathogen	MIC (µg/mL)
E. coli	16
B. subtilis	2
S. aureus	4
C. albicans	64

Source: Su (2011).

2.1.52   CATFISH (ORDER: SILURIFORMES; FAMILY: CLARIIDAE)

Clarias gariepinus (Burchell, 1822)

Source: Image by W.A. Djatmiko. https://creativecommons.org/licenses/by-sa/3.0/

Common name(s): North African catfish

Global distribution: This subtropical species is commonly seen in Lebanon, Israel, Syria, Jordan, and Turkey. However, it has been introduced to Africa, Europe, and Asia

Zone of inhibition (mm dia.).

	Ethanol		Methanol		Butanol		Acetone		Chloroform	
	Skin	Spine	Skin	Spine	Skin	Spine	Skin	Spine	Skin	Spine
Bacterial pathogens										
Streptococcus pneumoniae	13	3	2	2	3	4	6	6	2	13
Staphylococcus aureus	3	6	2.5	8	5	5	3	3	2	2
Vibrio cholera	3	2	5	5	4	10	2	3	5	3
Salmonella typhi	3	3	4	3	3	3	4	4	3	2
Klebsiella pneumoniae	10	4	12	8	22	11	19	13	20	13
Escherichia coli	3	3	4	5	4	8	3	10	4	14
Bacillus subtilis	13	10	8	10	5	9	5	18	3	2
Fungal pathogens										
A. fumigatus	5	8	4	8	10	2	2	3	16	3
Rhizomucor mehei	8	8	7	7	9	5	2	7	3	2
Candida tropicalis	12	10	9	5	7	5	2	5	2	13
Candida glabrata	15	4	3	2	12	3	3	2	2	5
Candida albicans	9	10	5	10	8	9	15	10	12	3

Source: John et al. (2015).

Habitat: It occurs mainly in lakes, rivers floodplains, swamps, and man ponds and its depth range is 0–80 m. It has accessory air-breathing organs to help its survival in shallow mud between rainy seasons.

Maximum length and weight: 170 cm; 60.0 kg

Food and feeding: It is a nocturnal fish feeding on living and dead animal matter. It is also able to pry on water birds.

Uses: This species has minor commercial fisheries. It is also an aquaculture species with major economic importance.

Pharmaceutical and nutraceutical compounds and activities

Antimicrobial properties: The slime of this fish has been reported to possess antibacterial activity on *Escherichia coli*, *Pseudomona aeruginosa*, *Staphylococcus aureus*, *Bacillus subtilis*, *Salmonella typhi*, and *Klebsiella pneumoniae*, and antifungal activity with *Candida albicans* and *Aspergillus niger* (Akunne et al., 2016).

Wound healing properties: The mucin ointment from this fish has been reported to possess high tensile strength and exhibit significant wound healing properties (Akunne et al., 2016).

2.1.53   CATFISH (ORDER: SILURIFORMES; FAMILY: HETEROPNEUSTIDAE)

Heteropneustes fossilis (Bloch, 1794)

Source: Francis Day and Suzini - Day, Francis (1878) The Fishes of India. Volume 2. Public domain.

Common name(s): Stinging catfish

Global distribution: Tropical Asian countries, such as Sri Lanka to Myanmar and Pakistan

Habitat: Adult fishes live in ponds. But they are also seen in swamps, marshes, ditches, and muddy rivers at a depth of about 1 m. It is able to thrive in brackish waters and rice fields.

Maximum length and weight: 31.0 cm

Food and feeding: It is an omnivorous species and its major preferred food items include insect larvae, insects, ostracods, gastropods, and plant materials.

Uses: This species has commercial fisheries with aquaculture and aquarium values.

Pharmaceutical and nutraceutical compounds and activities

Rajani and Alka (2015) in their studies on the ethno-medicinal importance of food fish, reported on the antimicrobial and anticancer properties of this catfish.

2.1.54   CATFISH (ORDER: SILURIFORMES; FAMILY: PANGASIIDAE)

Pangasianodon hypophthalmus **(Sauvage, 1878)**

Common name(s): Striped catfish

Global distribution: It is commonly distributed in Asian countries such as Mekong, Chao, and Phraya.

Habitat: It is a riverine freshwater species. As a potamodromous fish, it undergoes long-distance migrations between upstream spawning habitats and downstream feeding habitats.

Maximum length and weight: 130 cm; 44 kg

Food and feeding: It is an omnivore feeding mainly on crustaceans and fish. It may also feed on vegetable debris.

Uses: This species has commercial fisheries with aquaculture and aquarium values.

Pharmaceutical and nutraceutical compounds and activities

Antioxidant activity: The papain and bromelain protein hydrolysates of this species exhibited 90% DPPH radical scavenging activity (90%) and metal-chelating capacity (less than 20%). The reducing capacity of these hydrolysates was found to be 0.612 and 0.728 nm, respectively (Tanuja et al., 2012).

2.1.55   EELTAIL CATFISHES (ORDER: SILURIFORMES; FAMILY: PLOTOSIDAE)

Plotosus lineatus (Thunberg, 1787)

Common name(s): Striped eel catfish

Global distribution: Tropical; Indian Ocean, western Pacific Ocean; East Africa and Madagascar

Habitat: This benthic species is commonly seen in coral reefs at a depth range of 1–60 m. But it may also be found in open coasts, estuaries, and tide pools. It is amphidromous making regular migrations between freshwater and sea.

Maximum length: 32.0 cm

Food and feeding: Adults search and stir the sand incessantly for their food items, namely, crustaceans, mollusks, worms, and fish

Source: Image by Stan Shebs. https://creativecommons.org/licenses/by-sa/3.0/deed.en

Uses: Commercial fisheries exist for this species. It is also a commercially important aquarium fish.

Pharmaceutical and nutraceutical compounds and activities

Antimicrobial activity: The acetone, methanol, *n*-butanol, ethanol, and chloroform extracts of its skin and spine displayed antimicrobial activity (John et al., 2015). The inhibition zone values recorded on the different pathogens are given below.

Hemolytic activity: The spine-venom extracts of this species showed hemolytic activity in the erythrocytes of sheep, goat, human, and chicken with values of 7, 4, 7, and 10 HU, respectively (Pachaiyappan et al., 2015).

Cytotoxicity: The spine-venom extracts of this species showed cytotoxicity against HEp2 (epithelial cells derived from larynx carcinoma) with concentration-dependent cell death (Pachaiyappan et al., 2015).

2.1.56   SEAHORSES, PIPEFISHES (ORDER: SYNGNATHIFORMES; FAMILY: SYNGNATHIDAE)

Hippocampus kelloggi Jordan & Snyder, 1901

Source: Image by opencage.
https://creativecommons.org/licenses/by-sa/2.5/deed.en

Common name(s): Great seahorse

Global distribution: Red Sea to Japan; Australia and Africa

Habitat: This bathydemersal, nonmigratory marine species is found in deeper waters and is associated with coral reefs.

Maximum length: 28 cm

Food and feeding: Its food items include nematodes, copepods, amphipods, mysids, and tanaids; *Acetes* sp. and Lucifer; and shrimp larvae, crab larvae, and fish larvae.

Uses: It is an important aquarium species and a minimum size of 10 cm is employed in its trade.

Hippocampus kuda Bleeker, 1852

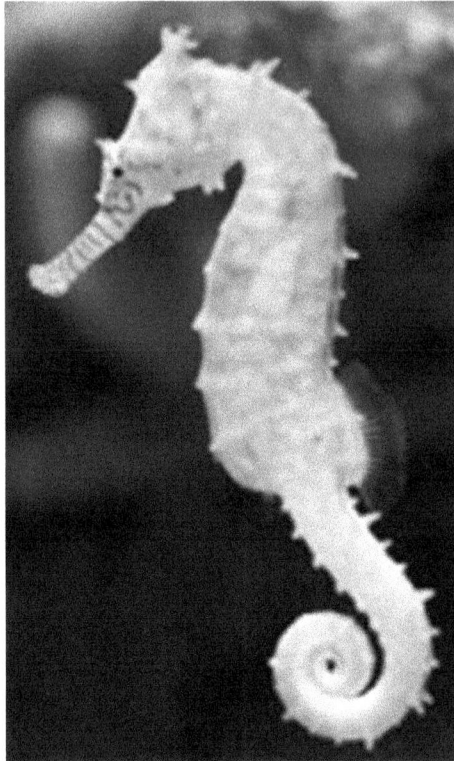

Common name(s): Spotted seahorse

Global distribution: Tropical Indo-Pacific regions such as India to southern Japan and Pakistan; Hawaii and Society Islands

Habitat: It is a reef-associated species with a depth range of 0–68 m. It is also seen in open water, steep mud slopes, and estuaries. Adults which are in pairs and non-migratory and are often seen in seagrass and marine algal areas.

Maximum length: 30.0 cm

Food and feeding: It feeds chiefly on zooplankton

Uses: It is an aquarium species and a minimum size of 10 cm is employed in its trade.

Hippocampus trimaculatus Leach, 1814

Source: C. Achilles. Day, Francis (1878) The Fishes of India. Volume 2. Public domain.

Common name(s): Longnose seahorse

Global distribution: Tropical; Indo-Pacific regions such as Australia and Tahiti; and India to Japan

Habitat: It is a nonmigratory species living in the gravel or sand bottoms near shallow reefs at a depth range of 0–100 m. It is also able to tolerate muddy estuaries and mangroves.

Maximum length: 22.0 cm

Food and feeding: It feeds mostly on crustaceans such as small caridean shrimps and amphipods

Uses: It is an aquarium species and a minimum size of 10 cm is employed in its trade. Commercial aquaculture exists for this species.

Pharmaceutical and nutraceutical compounds and activities of *Hippocampus spp.*

Antibacterial activity: The *n*-butanol extracts of the aforesaid species exhibited antibacterial activity on *Klebsiella pneumoniae, P. mirabilis, S. typhi, E. coli, S. aureus, V. parahemolyticus, S. aureus, S. paratyphi, K. oxytoca,* and *V. cholerae* and the zone of inhibition values are given below (Kumaravel et al., 2010).

Antimicrobial activity of *n*-butanol extracts of *Hippocampus* sp. (values of zone of inhibition (mm dia.)).

Pathogen	H. trimaculatus	H. kuda	H. kellogi
K. pneumoniae	7.5	0.1	4
P. mirabilis	0.1	0.1	0.1
S. typhi	0.1	4	0.1
E. coli	0.1	4	0.1
S. aureus	0.1	6	0.1
V. parahemolyticus	4	0.1	0.1
S. aureus	4	0.1	0.1
V. cholerae	7	0.1	0.1
S. paratyphi	0.1	5	4
K. oxytoca	4	6	0.1

Source: Kumaravel et al. (2010).

Antifungal activity: The methanol and butanol extracts of *H. kuda* exhibited antifungal activity on *Trichophyton mentagrophytes* and *Aspergillus flavus* with inhibition zone values of 4 and 2 mm, respectively (Kumaravel et al., 2010).

Neuroprotective activity: The peptides HTP-1 derived from the protein hydrolysates of *H. trimaculatus* exhibited neuroprotective activity with significant protective effects against neuronal cells death in Alzheimer's disease in-vitro model (Pangestuti and Kim, 2015).

Syngnathus acus Linnaeus, 1758

Source: Image by Gervais et Boulart. Public domain.

Common name(s): Greater pipefish

Global distribution: Subtropical; Eastern Atlantic regions such as British Isles and Mediterranean Sea

Habitat: It is a nonmigratory species living on sand, mud, and rough bottoms of coastal and estuarine waters to depths of at least 110 m. It is also found associated with marine algae and eel grass (Zostera)

Maximum length: 50.0 cm

Food and feeding: Feeds mainly on small crustaceans

Uses: It has no fisheries value.

Pharmaceutical and nutraceutical compounds and activities

Antitumor activity: The protein syngnathusin derived from the whole body of this species exhibited antitumor activity by inhibiting the growth of S180, CCRF-CEM, and A54 (Khora, 2013).

Pangestuti and Kim (2017) reported that syngnathusin was pro-apoptotic on A549 (IC_{50}: 84.9 µg/mL) and CCRF-CEM (IC_{50}: 215.3 µg/mL) cells.

Syngnathus schlegeli Kaup, 1856

Source: Image courtesy of Andrew Cornish

Common name(s): seaweed pipefish

Global distribution: Tropical; Northwest Pacific regions such as Vladivostok and Gulf of Tonkin

Habitat: This benthic species is found both in the marine and brackish water environments including estuaries. It is often found associated with seagrass, *Zostera*, and other marine algae. It is oceanodromous.

Maximum length: 30.0 cm

Food and feeding: It is a carnivore feeding mainly on copepods, gammarid amphipods, tanaids, and mysids. It consumes copepods in its initial feeding stage during spring; and gammarid amphipods with the increment of the size in summer and fall (Sung-Hoi, 1997).

Uses: It is not commercially important and it may have little aquarium value.

Pharmaceutical and nutraceutical compounds and activities

ACE-I inhibitory activity: The alcalase hydrolysate of the muscle protein of this species showed ACE-I inhibitory activity. Two fractions of this hydrolysate had amino acid sequences of Thr–Phe–Pro–His–Gly–Pro and His–Trp–Thr–Thr–Gln–Arg, respectively, and showed ACE-I inhibition with IC_{50} values of 0.62 mg/mL and 1.44 mg/mL, respectively (Wijesekara et al., 2011).

2.1.57   TRIGGERFISH (ORDER: TETRAODONTIFORMES; FAMILY: BALISTIDAE)

Balistes capriscus Gmelin, 1789

Common name(s): Gray triggerfish

Global distribution: Western Atlantic: Nova Scotia to Argentina; eastern Atlantic: Mediterranean Sea and Angola

Habitat: It dwells in coral reefs, harbors, lagoons, and bays at a depth range of 0–100 m. It may form schools and may be associated with seaweeds like *Sargassum* sp.

Maximum length and weight: 60.0 cm; 6.2 kg

Food and feeding: its main food items include benthic mollusks and crustaceans.

Uses: This triggerfish is a commercially and recreationally important fish. This species has commercial fisheries. Its flesh is a delicacy and is eaten fresh, smoked, and dried/salted. Caution is however to be exercised due to its ciguatera poisoning. It is also a valuable aquarium fish.

Pharmaceutical and nutraceutical compounds and activities

Antioxidant activity: The gelatin derived from this species has been reported to exhibit antioxidant activity (Jridia et al., 2019).

Nutraceutical properties: High contents of protein (20%) and amino acids (18.5%) especially leucin and lysine make this species a functional food for people with malnutrition (Levinton et al., 1981).

2.1.58   PORCUPINEFISH, BLOWFISH (ORDER: TETRAODONTIFORMES; FAMILY: DIODONTIDAE)

Cyclichthys orbicularis (Bloch, 1785)

Source: Image by Bernard Dupont from France. https://creativecommons.org/licenses/by-sa/2.0/deed.en

Common name(s): Birdbeak burrfish

Global distribution: Temperate Australasia and Africa; and Tropical Indo-Pacific

Habitat: It is a marine species inhabiting sand and mud bottoms in clear protected reefs where with seaweed and sponges are abundant. It has a depth range of 9–170 m.

Maximum length: 30.0 cm

Food and feeding: It is a nocturnal species presumably feeding on hard-shelled invertebrates.

Uses: Subsistence fisheries exist for this game fish.

Pharmaceutical and nutraceutical compounds and activities

Antibacterial activity: A maximum antibacterial activity (inhibition zone, 6 mm dia.) on *Proteus mirabilis*; and minimum activity (3 mm dia.) on *Klebsiella oxytoca*, *Vibrio cholerae*, and *Vibrio parahaemolyticus* have been recorded with the crude methonol extracts of this species (Raj et al., 2015).

Diodon holocanthus **Linnaeus, 1758**

Source: Image by Strobilomyces. https://creativecommons.org/licenses/by-sa/3.0/

Common name(s): Longspined porcupinefish

Global distribution: Throughout the subtropical world's oceans; western Atlantic Ocean: Florida and Bahamas to Brazil; Eastern Atlantic; and eastern and central Pacific

Habitat: It is a benthopelagic species dwelling on the rocky and soft bottoms near reefs at a depth range of 2–200 m. While young and subadults are small schoolers and juveniles are solitary with floating *Sargassum* rafts.

Maximum length and weight: 50.0 cm

Food and feeding: The main food items of this nocturnal feeder include hermit crabs, brachyuran crabs, mollusks, and sea urchins.

Uses: This species has minor commercial fisheries with high aquarium value.

Pharmaceutical and nutraceutical compounds and activities

Antibacterial activity: The crude methanol extracts of this species showed maximum antibacterial activity on *Escherichia coli* and minimum activity with *Proteus mirabilis* with inhibition zone values of 6 and 1 mm, respectively (Raj et al., 2015).

2.1.59   LEATHERJACKETS (ORDER: TETRAODONTIFORMES; FAMILY: MONACANTHIDAE)

***Thamnaconus septentrionalis* (Günther, 1874)** (=*Navodon septentrionalis*)

Source: http://www.fishbiosystem.ru/TETRAODONTIFORMES/Monacanthidae/Thamnaconus_septentrionalis2.html

Common name(s): Bluefin leatherjacket

Global distribution: Temperate; Indo-West Pacific: Korean Peninsula, Japan, and China Sea to Africa

Habitat: This marine, benthic species is living at a depth range of 50–120 m

Maximum length: 23.0 cm

Food and feeding: Its food items include zooplankton, namely, copepods, amphipods, and ostracods, and benthic mollusks.

Uses: It is a commercially important food fish and is sold as frozen seafood. It is also used in Chinese medicine.

Pharmaceutical and nutraceutical compounds and activities

Antioxidant activity: Three peptides with the amino acid sequence of as Gly–Ser–Gly–Gly–Leu (BSP-A), Gly–Pro–Gly–Gly–Phe–Ile (BSP-B), and Phe–Ile–Gly–Pro (BSP-C), respectively, have been derived from the alcalse hydrolysate of this species. Among these peptides, BSP-C showed maximum DPPH, HO, and O_2 scavenging activities with EC50 values of 0.118, 0.073, and 0.311 mg/mL, respectively (Chi et al., 2015b).

Chi et al. (2015a) reported that three peptides with the amino acid sequence of Trp–Glu–Gly–Pro–Lys (1), Gly–Pro–Pro (2), and Gly–Val–Pro–Leu–Thr (3) obtained from the head protein of this species exhibited antioxidant activity. Among these peptides, peptide 2 showed the maximum DPPH, hydroxyl, and ABTS radical scavenging activities with EC50 values of 1.927, 2.358, and 2.472 mg/mL, respectively. On the other hand, peptide 3 showed significant superoxide radical scavenging activity with an EC50 value of 2.881 mg/mL; and peptide 1 strongly inhibited the peroxidation of linoleic acid.

2.1.60   PUFFERFISH, TOADFISH (ORDER: TETRAODONTIFORMES; FAMILY: TETRAODONTIDAE)

Arothron hispidus (Linnaeus, 1758)

Source: Image by Factumquintus. https://creativecommons.org/licenses/by-sa/3.0/deed.en

Common name(s): White-spotted puffer

Global distribution: Throughout Indo-Pacific from Red Sea to Gulf of California and Panama

Habitat: Adults of this species live in coral reefs and rocky bottoms in a depth range of 1–50 m, and young ones are predominantly seen in estuaries. It is a solitary puffer.

Maximum length and weight: 50.0 cm; 2.0 kg

Food and feeding: Its food items include a wide range of invertebrates.

Uses: It has minor commercial fisheries owing to its most poisonous nature. It is a common species for marine aquarium.

Pharmaceutical and nutraceutical compounds and activities

Larvicidal (antimalarial) activity: The liver, ovary, skin, and muscle extracts of this pufferfish showed larvicidal activity against mosquitoes, namely, *Anopheles stephensi*, *Culex quinquefasciatus*, and *Aedes aegypti* and LC50 values are given below.

Larvicidal activity (LC50 (ppm)) of *A. hispidus* tissue extracts on mosquito larvae.

Mosquito	Liver	Ovary	Skin	Muscle
A. stephensi	1194.3	1421.4	7116.9	10,817.8
C. quinquefasciatus	1382.7	1982.7	15,039	ND

ND: No data.

Source: Samidurai and Mathew (2013).

Antimicrobial and cytotoxic activities: The skin extract of this species has been reported to exhibit significant antimicrobial activity on *E. coli* and *A. niger*. On the other hand, its liver extract exerted weak activity against the bacterial strain *Proteus vulgaris* and fungus, *Trichoderma viridae*. The skin extract of this species also showed cytotoxic activity on HeLa 2 cell line with an IC_{50} value of 1.78 µg/mL (Priya and Khora, 2013).

Hemolytic activity: Methanol extracts of this species have shown cytolytic activity on human blood RBCs and its hemolytic titer value was found to be 32 (Raj et al., 2015).

Arothron immaculatus (Bloch & Schneider, 1801)

Source: Image by Philippe Bourjon. https://creativecommons.org/licenses/by-sa/3.0/

Order: Tetraodontiformes; Family: Tetraodontidae

Common name(s): Immaculate puffer

Global distribution: Tropical Indo-West Pacific regions such as Australia, Indonesia, Madagascar, and Africa

Habitat: This reef-associated species dwells in marine and brackish water areas; depth range 3–30 m; it is solitary and is also common in weedy areas, estuaries, seagrass beds, and mangrove areas.

Maximum length: 30.0 cm

Food and feeding: It feeds mainly on benthic crustaceans and mollusks although they are also known to feed on seagrass and mangrove plants.

Uses: It has minor commercial fisheries only for public aquariums owing to its poisonous nature.

Pharmaceutical and nutraceutical compounds and activities

Antibacterial activity: The chloroform extract of the liver of this species showed antibacterial activity against *Staphylococcus aureus* with the maximum zone of inhibition (2.5 mm). On the other hand, the ethanolic extract of liver acted against *V. parahemolyticus* and *V. cholerae* with the value of 2.0 mm. The minimum zone of inhibition (0.5 mm) was against *P. mirabilis* in the ethanolic extract. In regards, the skin extracts of this species, maximum zone of inhibition was found against *V. cholerae* of 9.8 mm in methanolic extract; 5.7 mm in *V. parahemolyticus* with ethanolic extract; and a minimum of 0.2 mm in *S. typhi* with chloroform extract. The MIC value of liver extract against *S. aureus* was recorded as 150 L and MIC of skin extract against *V. cholerae* was found to be 260 L (Kumaravel et al., 2011).

Antidiabetic and antioxidant activities: The muscle methanol extract of this species had higher efficiency to scavenge the free radicals. Further, this extract showed activity against the high-fat fed and streptozotocin-induced diabetic experimental rats by significantly reducing the blood glucose (Kaleshkumar et al., 2019).

Arothron stellatus (Bloch & Schneider, 1801)

Source: Image by Richard Ling. https://creativecommons.org/licenses/by-sa/2.0/

Common name(s): Stellate puffer

Global distribution: Tropical Indo-Pacific regions such as Red Sea and East Africa; Indonesia to Tuamotus; Japan; and Lord Howe Island

Habitat: This reef-associated, diurnal species dwells in marine and brackish water areas at depths of 3–58 m. Juveniles are seen inshore, usually on muddy substrates and often estuarine.

Maximum length: 120 cm

Food and feeding: Its main food items are benthic invertebrates.

Uses: There are no fisheries for this species. As it is a poisonous species.

Pharmaceutical and nutraceutical compounds and activities

Antibacterial, antifungal, and cytotoxic activities: The crude extracts of this species showed antibacterial activity against *E. coli*, *S. aureus*, *B. cereus*, *B. subtilis*, *K. pneumoniae*, *P. vulgaris*, and *P. vulgaris*; and antifungal activity against *A. niger*, *A. flavus*, *A. fumigatus*, *C. albicans*, *Trichoderma viride*, and *T. rubrum*. Further, these extracts showed cytotoxic activity on HeLa cell line. The values recorded with antimicrobial activity and cytotoxicity are given in the following tables

Antibacterial activity of crude extract.

	Zone of inhibition (mm dia.)
B. cereus	10.2
K. pneumoniae	8.2
P. aeruginosa	9.1
E. coli	11.3
S. aureus	11.1
P. vulgaris	10.4

Antifungal activity of crude extract.

	Zone of inhibition (mm dia.)
T. viride	8.0
T. rubrum	8.1
A. niger	10.0
A. flavus	10.5
A. fumigatus	10.2
C. albicans	9.5

Cytotoxic effect of crude extract on HeLa cell line.

Concentration (μg/mL)	% of inhibition
3	5.1
6	11.2
12.5	19.3
25	24.8
50	41.3
100	63.2

Source: Jal et al. (2014).

Antibacterial activity: Raj et al. (2015) reported that the crude methanol extracts of this species showed maximum antibacterial activity on *Streptococcus pyogens* and minimum activity against *Salmonella typhi* with inhibition zone values of 8 and 2 mm, respectively.

Canthigaster solandri **(Richardson, 1845)**

Source: Image by Tànsu k nd ng. https://creativecommons.org/licenses/by-sa/2.0/deed.en

Common name(s): Spotted sharpnose

Global distribution: Tropical Indo-Pacific: East Africa to Tuamotu; Ryukyu Islands, New Caledonia and Tonga; Hawaiian Islands

Habitat: This marine, benthopelagic species lives in rocky reefs, reef flats, and lagoons at a depth range of 10–36 m. This species may swim in pairs or groups.

Maximum length: 11.5 cm

Food and feeding: Its major food items include green and red algae; corals, bryozoans, polychaetes, crustaceans, mollusks, echinoderms, and tunicates.

Uses: This species has commercial aquarium value.

Pharmaceutical and nutraceutical compounds and activities

Antibacterial activity: The crude methanol extracts of this species showed maximum antibacterial activity on *Salmonella paratyphi* and minimum activity on *Klebsiella pneumoniae* with inhibition zone diameter values of 7 and 2 mm, respectively (Raj et al., 2015).

Chelonodon patoca (Hamilton, 1822)

Source: Image by Totti, https://creativecommons.org/licenses/by-sa/4.0/

Common name(s): Milk-spotted puffer

Global distribution: Tropical Indo-Pacific areas such as India, Australia, China, and East Africa

Habitat: This reef-associated, anadromous species lives usually in sand and mudflats of marine and brackish water areas at a depth range of 4–60 m. It may also enter freshwater, estuaries, and mangroves.

Maximum length and weight: 38.0 cm; 10.2 kg

Food and feeding: It is a carnivorous species preying mainly on mollusks and worms

Uses: This species has only minor commercial fisheries. In Japan, it is a delicacy.

Pharmaceutical and nutraceutical compounds and activities

Larvicidal (antimalarial) activity: The liver extracts of fish displayed larvicidal activity on the larvae of mosquitoes, namely, *Anopheles stephensi*, *Culex quinquefasciatus*, and *Aedes aegypti* and LC50 values recorded were found to be 1182.3, 1543.0, and 2441.0 ppm, respectively (Samidurai and Mathew, 2013).

Dichotomyctere fluviatilis (Hamilton, 1822) (*=Tetraodon fluviatilis*)

Source: Image by Steven G. Johnson. https://creativecommons.org/licenses/by-sa/3.0/

Common name(s): Green pufferfish

Global distribution: Tropical Asian countries such as Sri Lanka, India, Bangladesh, and Myanmar

Habitat: This benthic species lives in freshwater and brackish water areas. It is potamodromous and is commonly seen in estuaries and slow-moving rivers.

Maximum length and weight: 17.0 cm

Food and feeding: Its main food items include crustaceans, mollusks, and other invertebrates; and vascular plants and detritus. It is an occasional feeder on fish scales and fins.

Uses: As the muscular tissue and viscera of this species are extremely toxic, it has no commercial fisheries. Though it has some aquarium value, caution is to be exercised on adults which are pugnacious and aggressive with their tank mates.

Pharmaceutical and nutraceutical compounds and activities

Antibacterial activity: The crude extracts of the whole body of this species displayed antibacterial activity against pathogenic bacteria, namely, *Staphylococcus aureus*, *Streptococcus pyogenes*, *Pseudomonas aeruginosa*, *Klebsiella pneumoniae*, *Salmonella typhi*, and nonpathogenic bacteria *Escherichia coli* with the inhibition zone values ranging from 6 to 20 mm. The maximum and minimum values were, however, associated with *Escherichia coli* and *Salmonella typhi*, respectively. Among the different organ extracts, the liver extracts registered the maximum value (20 mm) (Zodape, 2018).

Lagocephalua inermis (Temminck & Schlegel, 1850)

Source: Image by Kawahara Keiga. Public domain.

Common name(s): Smooth blaasop

Global distribution: Tropical Indo-West Pacific regions, namely, South Africa to southern Japan

Habitat: This marine, demersal species is largely found in the continental shelf edge at a depth range of 10–200 m.

Maximum length: 90.0 cm

Food and feeding: It is a carnivorous species.

Uses: It has low economic value as it is a trash fish.

Pharmaceutical and nutraceutical compounds and activities

Larvicidal (antimalarial) activity: The extracts of liver, ovary, skin, and muscle tissues of this pufferfish showed larvicidal activity against species of mosquitoes such as *Aedes aegypti*, *Anopheles stephensi*, and *Culex quinquefasciatus*, and the record LC50 values against each species are given below.

Larvicidal activity of *L. inermis* tissue extracts against mosquito species.

Tissue	Mosquito	LC50 (ppm)
Skin	*A. stephensi*	10,283.0
Muscle	*A. stephensi*	6067.5
Liver	*C. quinquefasciatus*	1556.1
	A. aegypti	2426.4
Ovary	*A. stephensi*	1653.5
	C. quinquefasciatus	2734.7

Source: Samidurai and Mathew (2013).

Hemolytic activity: The crude extracts of this species showed activity with an hemolytic titer value of 16 against human erythrocytes (Raj et al., 2015a).

Antibacterial activity: The crude methonol extracts of this species showed maximum antibacterial activity on *S. paratyphi* and *V. cholerae* with a maximum inhibition zone diameter value of against maximum activity (6 mm) against *S. paratyphi* and *V. cholerae* (inhibition zone dm, 6 mm) and minimum activity (1 mm) showed against *Proteus mirabilis* (Raj et al., 2015)

Lagocephalus scleratus (Gmelin, 1789)

Source: Image by Rickard Zerpe. https://creativecommons.org/licenses/by/2.0/

Common name(s): Silver-cheeked toadfish

Global distribution: Tropical Indo-West Pacific Ocean, Red Sea, and Mediterranean Sea

Habitat: This species inhabits offshore reefs with sandy bottoms at a depth range of 18–100 m.

Maximum length and weight: 110 cm; 7.0 kg

Food and feeding: It is carnivorous fish feeding mainly on shrimps and crabs; squids and cuttlefish; and fish including its own species.

Uses: It is only a commercial aquarium species.

Pharmaceutical and nutraceutical compounds and activities

Larvicidal (antimalarial) activity: The skin, muscle, liver, and ovary extracts of this fish showed larvicidal activity on species of mosquitoes such as *Culex quinquefasciatus*, *Anopheles stephensi*, and *Aedes aegypti*, and the recoded LC50 values on these species are given below.

Larvicidal activity of *Lagocephalus scleratus* extracts against *A. stephensi*, *C. quinquefasciatus*, and *A. aegypti*.

Tissue	Mosquito	LC50 (ppm)
Ovary	*A. stephensi*	1414.9
	C. quinquefasciatus	2278.7
Liver	*A. stephensi*	1510.0
	C. quinquefasciatus	1608.7

Source: Samidurai and Mathew (2013).

Antimicrobial activity: The crude tetrodotoxin extracts of liver, skin, and muscles of this pufferfish showed antimicrobial activity (Alabssawy, 2017). The inhibition zone diameter values obtained for these extracts against the different species of pathogens are given in the following table.

Antimicrobial activity of TTX extracts of liver, skin, and muscles of *L. sceleratus*.

	Liver	Skin	Muscle
	IZD (mm)	IZD (mm)	IZD (mm)
Bacterial species			
Escherichia coli	22.2	20.2	19.2
Salmonella typhi	12.1	ND	12.7
Staphylococcus aureus	13.8	12.8	11.3
Bacillus subtilis	17.5	16.3	15.3
Streptococcus agalactiae	12.7	10.8	11.4
Vibrio cholerae	9.2	8.5	8.1
Aeromonas veronii	20.5	18.0	19.7
Fungal species			
Aspergillus fumigatus	19.2	17.1	16.3
Trichophyton rubrum	ND	5.3	6.2
Candida albicans	12.5	13.2	10.7

IZD: Inhibition zone diameter.; ND: No data.

Source: Alabssawy (2017).

Lagocephalus spadiceus (**Richardson, 1845**) (*=Gastrophysus spadiceus*)

Source: Image by agancz. https://creativecommons.org/publicdomain/zero/1.0/deed.en

Common name(s): Half-smooth golden pufferfish

Global distribution: Subtropical Indo-West Pacific countries such as South Africa and Australia

Habitat: This demersal, oceanodromous species is occasionally enters the mouths of rivers (estuaries).

Maximum length: 37.4 cm

Food and feeding: It is a carnivore species feeding largely on benthic teleosts (*Upeneus guttatus*), crustaceans (*Aristeus* sp.).

Uses: It has no food value as it is a trash fish.

Pharmaceutical and nutraceutical compounds and activities

Antibacterial activity: The skin extract of this species exhibited maximum activity against *V. cholera* and minimum activity with *S. pyogenus* with inhibition zone values of 24 and 14 mm, respectively. On the other hand, its muscle extract showed the maximum (22.4 mm) and minimum (16.2 mm) values with *V. parahaemolyticus* and *B. subtilis*, respectively (Priya et al., 2016).

Antifungal activity: The antifungal activity of the skin extract showed the highest zone of inhibition against *A. terreus* (19.8 mm) and the minimum activity was observed against *A. fumigatus* (11.7 mm). On the other hand, the muscle extract registered the maximum zone of inhibition with *A. terreus* (18.9 mm) and minimum activity with *A. flavus* (13.5 mm). In liver extract, the highest zone of inhibition was observed against *A. niger* (18.5 mm) and the minimum activity against *C. albicans* (11.6 mm) (Priya et al., 2016).

Takifugu oblongus (Bloch, 1786) (=*Tetrodan oblongus*)

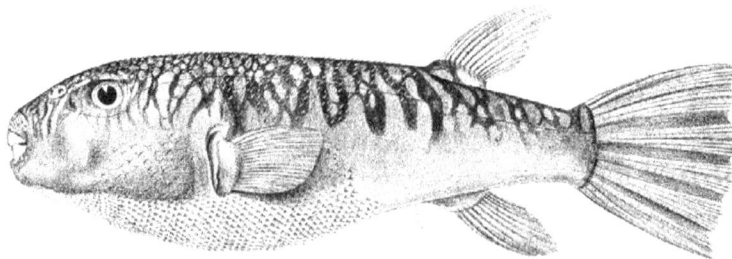

Source: Sir Francis Day (1878) The Fishes of India. Volume 2. Public domain.

Common name(s): Lattice blaasop

Global distribution: Tropical Indo-West Pacific countries, namely, South Africa to Indonesia; and Japan and Australia

Habitat: This demersal species lives in shallow coastal waters at a depth range of 0–20 m and occasionally it enters brackish waters.

Maximum length: 40.0 cm

Food and feeding: Its diet consists mainly of algae, mollusks, and invertebrates including crustaceans.

Uses: It has no fisheries value as it is a toxic fish.

Pharmaceutical and nutraceutical compounds and activities

Antibacterial activity: The skin extracts of this species showed maximum antibacterial activity on *B. subtilis* and *K. pneumoniae* with an inhibition zone dia. value of 17 mm and a minimum inhibition zone dia. of 7 mm was recorded with *P. aeruginosa*. On the other hand, the muscle extracts had a maximum inhibition zone dia. value of 11 mm with *S. aureus* and minimum value (10 mm) with *E. coli* and *K. pneumoniae*. This muscle extract, however, was found ineffective with *B. subtilis* and *P. aeruginosa*. In the liver extracts, maximum value (22 mm) was with *K. pneumoniae* followed by *E. coli* (18 mm). In the gonad extracts, maximum value (28 mm) was recorded on *S. aureus* and *K. pneumoniae* followed by *E. coli* and *P. aeruginosa* (24 mm) (Indumathi et al., 2016). The inhibition zone values of bacterial species registered with different tissue extracts are given in the following table.

Antibacterial activity (inhibition zone dia., mm) of the tissue extracts of *T. oblongus* on bacterial pathogens.

Tissue extracts/Bacterial pathogens	Concentration of extract (µL/well)	
	50	**100**
Skin		
E. coli	7.3	11.0
S. aureus	7	12.5
K. pneumoniae	12.2	17.2

Tissue extracts/Bacterial pathogens	Concentration of extract (µL/well)	
	50	100
B. subtilis	14.5	17.2
Muscle		
E. coli	8.2	10.1
S. aureus	8.2	11.0
K. pneumoniae	8.4	10.5
Liver		
E. coli	13.3	18.3
P. aeruginosa	9.6	12.0
S. aureus	9.5	14.5
K. pneumoniae	16.2	22.3
B. subtilis	8.1	10.5
Gonads		
E. coli	18.4	24.3
P. aeruginosa	18.4	24.1
S. aureus	18.5	28.1
K. pneumoniae	18.5	28.4
B. subtilis	7.5	9.3

Source: Indumathi et al. (2016).

Larvicidal activity: The organ extracts of this species showed larvicidal activity against mosquito larvae. Total activity (100% mortality) was observed at 40% concentration of all the extracts, while skin, liver, and gonads extracts showed highest mortality at 30% too. Skin extract showed exceptional larvicidal activity in all concentrations, producing 100% mortality of all the larvae. At 40% concentration, the muscle extract exhibited 100% mortality of the larvae. Liver and gonads extracts recorded the maximum, that is, 100% mortality at 30% and 40% concentrations, respectively. Interestingly, the less mortality of *Culex* sp. and *Anopheles* sp. was found with liver extract and gonads extract, respectively (Indumathi et al., 2016). The values of percentage of mortality of different larvae by various concentrations of the tissue extracts are given in the following table.

Larvicidal activity (%) of the tissue extracts of *T. oblongus* on mosquito larvae.

Tissue extracts/Mosquito sp.	Concentration (%)		
	10	20	30
Skin			
Culex sp.	100	100	100
Anopheles sp.	100	100	100
Aedes sp.	100	100	100
Muscle			
Culex sp.	18	52	97
Anopheles sp.	3	23	83
Aedes sp.	33	92	100
Liver			
Culex sp.	33	67	100
Anopheles sp.	57	98	100
Aedes sp.	78	100	100
Gonads			
Culex sp.	13	88	100
Anopheles sp.	37	65	100
Aedes sp.	38	82	100

Source: Indumathi et al. (2016).

2.2 CARTILAGINOUS FISHES (PHYLUM, CHORDATA; SUBPHYLUM, VERTEBRATA; CLASS: CHONDRICHTHYES)

2.2.1 REQUIEM SHARKS (ORDER: CARCHARHINIFORMES; FAMILY: CARCHARHINIDAE)

Carcharhinus acronotus (Poey, 1860)

Common name(s): Blacknose shark

Global distribution: Its distribution is limited to tropical and warm temperate western Atlantic Ocean.

Habitat: While adults of this species are living in the sandy and coral bottoms of continental shelves at a depth range of 9–64 m, juveniles are largely found in shallow water.

Maximum length and weight: 200 cm; 18.9 kg

Food and feeding: Its main food items include small fishes such as porcupine fish and pinfish.

Uses: Minor commercial fisheries exist for this game fish.

Pharmaceutical and nutraceutical compounds and activities

Antibacterial activity: Luer and Walsh (2018) reported on the antibacterial activity of its mucus-associated bacteria.

Conrad et al. (https://digitalcommons.northgeorgia.edu/gurc/2018/master-schedule/19/) also reported that the bacteria associated with this shark showed antibiotic activity against six human pathogenic bacterial test strains.

Carcharhinus leucas (Müller & Henle, 1839)

Source: Image by Chaloklum Diving. https://creativecommons.org/licenses/by/3.0/

Common name(s): Bull shark

Global distribution: Worldwide tropical to subtropical coastal waters; and Amazon River (Peru) and Mississippi River (Illinois)

Habitat: This species prefers shallow coastal waters of less than 30 m depth and its range is, however, 1–150 m. It is also common in estuaries, river mouths, bays, and lagoons. Juveniles are largely seen in estuaries and lagoons.

Maximum length and weight: 360 cm; 316.5 kg

Food and feeding: Its food items include other sharks, rays, bony fishes, mantis shrimps, crabs, sea snails, squid, sea urchins, sea turtles, and so on.

Uses: This species has commercial fisheries with game fish value.

Pharmaceutical and nutraceutical compounds and activities

Antibacterial activity: The mucus-associated bacteria of this species produced antibacterial activity (Luer and Walsh, 2018).

Conrad et al. (https://digitalcommons.northgeorgia.edu/gurc/2018/master-schedule/19/) also reported that the bacteria associated with this shark showed antibiotic activity against six human pathogenic bacterial test strains.

Carcharhinus limbatus (Müller & Henle, 1839)

Source: CSIRO National Fish Collection. https://creativecommons.org/licenses/by/3.0/

Common name(s): Blacktip shark

Global distribution: It is cosmopolitan in tropical to subtropical Atlantic; Gulf of Mexico and Caribbean Sea; Massachusetts to Brazil; Mediterranean and West Africa; Southern California to Peru; Galapagos Islands, Hawaii, Tahiti, and Australia; South Africa, Madagascar, Red Sea, Persian Gulf, Inai, and China

Habitat: It inhabits inshore and offshore waters; river mouths, estuaries, bays and mangrove swamps at a depth range of 0–100 m.

Maximum length and weight: 275 cm; 122.8 kg

Food and feeding: Its chief food items include pelagic and benthic fishes, small sharks and rays, and crustaceans and cephalopods. It is believed to be an active hunter in midwaters.

Uses: Commercial fisheries exist for this game fish.

Pharmaceutical and nutraceutical compounds and activities

Antibacterial activity: The mucus-associated bacteria of this species produced antibacterial activity (Luer and Walsh, 2018).

Conrad et al. (https://digitalcommons.northgeorgia.edu/gurc/2018/master-schedule/19/) also reported that the bacteria associated with this shark showed antibiotic activity against six human pathogenic bacterial test strains.

Others: The ASC and PSC derived from the cartilage of this species had significant amount of protein with traces of fat. Further, the presence of rich amino acids such as glycine, alanine, proline, and hydroxyproline make these collagens of this species suitable in the preparation of functional foods (Kittiphattanabawon et al., 2010).

Galeocerdo cuvier (Péron & Lesueur, 1822)

Source: Image by Albert Ko. https://creativecommons.org/licenses/by-sa/3.0/

Common name(s): Tiger shark, leopard shark

Global distribution: It is found distributed in world's warm temperate and tropical oceans.

Habitat: This species is normally found in shallow coastal waters and open ocean at a depth range of 0–800 m. It is also largely seen in lagoons, coral atolls, harbors, and estuaries. It undertakes seasonal migrations between temperate and tropical waters. It also makes long oceanic migrations and is capable of traveling long distances.

Maximum length and weight: 750 cm; 3110 kg

Food and feeding: Its chief food items include crustaceans, squid, sea turtles, rays, bony fishes, sea birds, dolphins, and so on. It is also believed to eat on carrion and garbage. It is a nocturnal feeder.

Uses: Commercial fisheries exist for this species. It is valued for its meat, fins, hide, liver oil, and also for its jaws and cartilage. It is also often used for fishmeal. For human consumption, it is sold fresh, dried-salted, smoked, and frozen. It also has sport fish value.

Pharmaceutical and nutraceutical compounds and activities

Antibiotic activities: The bacteria associated with this species showed antibiotic activities against one or more test strains (Conrad et al., https://digitalcommons.northgeorgia.edu/gurc/2018/masterschedule/19/).

Negaprion brevirostris (Poey, 1868)

Source: NMFS, E. Hoffmayer, S. Iglésias and R. McAuley. Public domain.

Common name(s): Lemon shark

Global distribution: Subtropical western Atlantic: New Jersey to Brazil; Gulf of Mexico and Caribbean; Senegal and Ivory Coast of Africa; North Pacific: Gulf of California and Baja California to Ecuador

Habitat: It normally inhabits shallow coastal waters at a depth range of 0–90 m; and coral reefs, bays, sounds, mangroves, and river mouths. Rarely, it migrates to oceanic regions. It may from small schools.

Maximum length and weight: 340 cm; 183.7 kg

Food and feeding: Its main food item is fish. It may also consume crustaceans and molluscs.

Uses: Commercial fisheries exist for this species. It is also a game fish.

Pharmaceutical and nutraceutical compounds and activities

Antibacterial activity: The mucus-associated bacterial isolates (10%) from this species showed antibacterial activity (Luer and Walsh, 2018).

Conrad et al. (https://digitalcommons.northgeorgia.edu/gurc/2018/masterschedule/19/) also reported that the bacteria associated with this shark showed antibiotic activity against six human pathogenic bacterial test strains.

Prionace glauca (Linnaeus, 1758)

Source: Image by Shane Anderson / U.S. National Oceanic and Atmospheric Administration. Public domain.

Common name(s): Blue shark

Global distribution: Tropical and temperate; Eastern Atlantic: Norway to South Africa and Mediterranean; Western Atlantic: Newfoundland to Argentina; central Atlantic; central Pacific; eastern Pacific: Gulf of Alaska to Chile; Indo-West Pacific: Arabian Sea to Indonesia; South Africa, Japan, Australia, and New Zealand

Habitat: This pelagic species lives in marine and brackish water at a depth range of 0–350 m and is often seen in the inshore areas around oceanic islands and narrow continental shelf.

Maximum length and weight: 400 cm; 205.9 kg

Food and feeding: Its main food items include teleost fishes, small sharks, crabs, squids, sea birds, and garbage.

Uses: This species has only minor commercial fisheries and it is a sport fish.

Pharmaceutical and nutraceutical compounds and activities

Anti-inflammatory activity: In Wistar rats, the chondroitin and glucos-amine extracts from the cartilage and the cartilage powder of this shark species showed inflammatory inhibition with the percentage values of 4.2, 3.9, and 4.0, respectively (Agustin et al., 2016).

2.2.2 CATSHARKS, DOGFISH (ORDER: CARCHARHINIFORMES; FAMILY: SCYLIORHINIDAE)

Scyliorhinus canicula (**Linnaeus, 1758**)

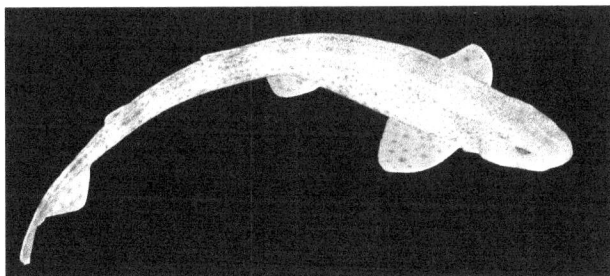

Source: Image by Hans Hillewaert. https://creativecommons.org/licenses/by-sa/4.0/

Common name(s): Small-spotted catshark, lesser spotted dogfish

Global distribution: Subtropical Northeast Atlantic: Britain and Ireland; Norway and Mediterranean

Habitat: It is commonly found in the sandy, muddy, coralline, gravel, and algal bottoms of the continental shelves and uppermost slopes at a depth range of 10–400 m. It is nocturnal preying actively at night and remaining calm during daytime. Juveniles are, however, found mainly in shallow water.

Maximum length and weight: 100.0 cm; 1.3 kg

Food and feeding: It feeds on benthic invertebrates such as mollusks, crustaceans, small cephalopods, polychaete worms, and small bony fishes. Males of this species have been found to forage in shallow areas with soft sediment or filamentous algae.

Uses: This species has only minor commercial fisheries and is utilized fresh and dried-salted. It is also utilized for the production of fishmeal and oil. It has also aquarium value.

Pharmaceutical and nutraceutical compounds and activities

ACE-inhibitory activity: Pangestuti and Kim (2017) reported that the peptides from the hydrolysates of this species showed ACE-inhibitory activity with an IC_{50} value of 0.44 μM.

Antihypertensive and antioxidant activities: The esperase hydrolysates of this species showed DPPH and ABTS radical scavenging activities, and antihypertensive activity (Vázquez et al., 2017).

2.2.3 HAMMERHEAD SHARKS (ORDER: CARCHARHINIFORMES; FAMILY: SPHYRNIDAE)

Sphyrna tiburo (Linnaeus, 1758)

Source: Image by D Ross Robertson. Public domain.

Common name(s): Bonnethead shark

Global distribution: Subtropical; eastern Pacific: California to Ecuador; western Atlantic: North Carolina (USA) to Brazil; Gulf of Mexico and Caribbean

Habitat: It is commonly seen in continental shelves at a depth range of 10-80 m; also in reefs, bays, and estuaries. It may form schools in groups of hundreds or thousands.

Maximum length and weight: 150 cm; 10.8 kg

Food and feeding: Its main food items include crustaceans, bivalves, mollusks, and other small fish.

Uses: This species has commercial fisheries with game fish value.

Pharmaceutical and nutraceutical compounds and activities

Antitumor activity: The compounds from the secretion of the epigonal cells [epigonal conditioned medium (ECM)] of this shark showed significant antitumor activity on B-cell lymphoma (Daudi), T-cell leukemia (Jurkat), melanoma (A375.S2), and fibrosarcoma (WEHI-164) cell lines; and weak activity against pancreatic cancer (PANC-1) and ovarian cancer (NIH: OVCAR-3) cell lines. Further, the above compounds had growth inhibition of malignant/nonmalignant cell line pairs, namely, Hs 578T/Hs 578Bst and HCC38/HCC38 BL (Walsh et al., 2006).

Cytotoxic activity: The ECM of this shark has been reported to demonstrate significant cytotoxic activity on T-cell leukemia cell line, Jurkat E6-1 cell lines by inducing apoptosis. EPM may also have the potential in the production of immune-regulatory compounds in the cancer therapy (Walsh et al., 2013; Luer and Walsh, 2018).

2.2.4 HOUND SHARKS (ORDER: CARCHARHINIFORMES; FAMILY: TRIAKIDAE)

Mustelus griseus Pietschmann, 1908

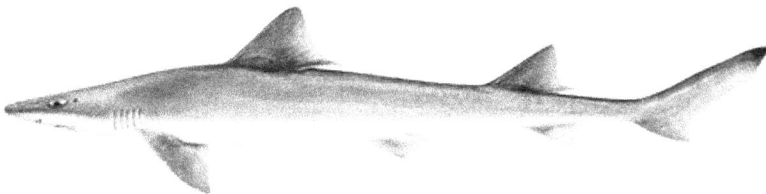

Common name(s): Spotless smooth hound

Global distribution: Tropical Northwest Pacific countries such as Korea, Japan, China, Taiwan, Vietnam, and the Philippines

Habitat: This marine, benthic species is common in inshore waters at a depth range of 5–300 m. It also dwells in semi-enclosed sea areas with sand bottom.

Maximum length: 87.0 cm

Food and feeding: Its main food items are invertebrates, especially benthic crustaceans.

Uses: This species has commercial fisheries. Its flesh and fins are highly esteemed as food items.

Pharmaceutical and nutraceutical compounds and activities

Antioxidant activity: The cartilage protein hydrolysate of this species yielded three peptides with the amino acid sequences of Gly–Ala–Glu–Arg–Pro (A), Gly–Glu–Arg–Glu–Ala–Asn–Val–Met (B), and Ala–Glu–Val–Gly (C), respectively. All these peptides have shown DPPH hydroxyl, ABTS, and superoxide anion radical scavenging activities with EC50 values of 3.73, 0.25, 0.10, and 0.09 mg/mL, respectively (A); 1.87, 0.34, 0.05, and 0.33 mg/mL, respectively (B); and 2.30, 0.06, 0.07, and 0.18 mg/mL, respectively (C) (Tao et al., 2018).

Mustelus mustelus (Linnaeus, 1758)

Source: Image by Gervais et Boulart. Public domain

Common name(s): Smooth hound

Global distribution: Temperate western South Atlantic: Brazil to Argentina

Habitat: This benthic species lives in the continental shelf at depths of 60–195 m.

Maximum length: 200 cm

Food and feeding: Its major food items include crustaceans, cephalopods, and bony fishes.

Uses: This species has commercial fisheries with game fish value.

Pharmaceutical and nutraceutical compounds and activities

Antioxidant and ACE-inhibitory activity: The hydrolysate derived from the intestine of this species exhibited DPPH radical scavenging activity with an IC_{50} value of 1.47 mg/mL. Further, the hydrolysate of alkaline protease extract (conc. 2 mg/mL) from the intestine showed 82% ACE-inhibitory activity. As this hydrolysate is rich in amino acids like methionine, phenylalanine, serine, valine, and leucine, this fish may have biotechnological and functional food applications (Sayari et al., 2016).

Satietogenic (appetite suppression) effect: A significant reduction in body weight and food intake has been recorded in experimental rats treated with the protein hydrolysate of this species. This finding suggests that this hydrolysate may be of great use in regulating appetite and preventing type II diabetes (Bougatef et al., 2010).

2.2.5 THRESHER SHARKS (ORDER: LAMNIFORMES; FAMILY: ALOPIIDAE)

Alopias pelagicus **Nakamura, 1935**

Source: NOAA Observer Program. Public domain.

Common name(s): Pelagic thresher

Global distribution: Temperate Pacific and Indian oceans; North America: California and Mexico; Taiwan

Habitat: Marine; pelagic-oceanic; oceanodromous (Ref. 51243); depth range 0–300 m. It is primarily an oceanic species but sometimes close inshore; neritic to oceanic (oceanodromous); it may enter atoll lagoons; and depth range 0–300 m.

Maximum length: 428 cm

Food and feeding: Stuns its prey with its tail and is presumably feeding on small fishes and cephalopods.

Uses: This species has commercial fisheries with game fish values.

Pharmaceutical and nutraceutical compounds and activities

ACE-inhibitory activity: The crude extracts of the pancreas and kidney of this species showed ACE-inhibitory activity with the percentage values of 55 and 46, respectively (Nomura et al., 2002). Pangestuti and Kim (2017) reported on the ACE-inhibitory activity of the muscle of this species with an IC_{50} value of 0.54 μM.

2.2.6   WHITE SHARKS (ORDER: LAMNIFORMES; FAMILY: LAMNIDAE)

Carcharodon carcharias (Linnaeus, 1758)

Common name(s): Great white shark

Global distribution: It is plenty in temperate seas and rare in tropical waters. Western Atlantic: Newfoundland to Florida; Gulf of Mexico, Bahamas and Cuba; Brazil to Argentina; eastern Atlantic: France to South Africa; Indian Ocean: Red Sea, to South Africa; Seychelles, Reunion and Mauritius; western Pacific: Siberia to New Zealand; central Pacific: Marshall and Hawaiian Islands; eastern Pacific: Panama to Chile; central Pacific: Alaska to Gulf of California.

Habitat: It is common in the nearshore waters at a depth range of 0–1200 m. It may also be found from the surfline to the open ocean. It may be seen singly or in pairs and it is not a schooler.

Maximum length and weight: 640 cm; 3400.0 kg

Food and feeding: Its food items are squids, crabs, octopi, sharks, rays, teleost fishes, seals, dolphins porpoises, whales, and sea birds.

Uses: This species has minor fisheries with game fish value.

Pharmaceutical and nutraceutical compounds and activities

Antibacterial activity: The bacterial isolates cultured from the mucus of this species exhibited antibacterial activity (Luer and Walsh, 2018).

2.2.7 WHIPTAIL STINGRAYS (ORDER: MYLIOBATIFORMES; FAMILY: DASYATIDAE)

Brevitrygon imbricata (**Bloch & Schneider, 1801**) (*=Himantura imbricata*)

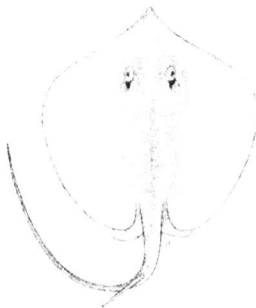

Common name(s): Bengal whipray

Global distribution: Tropical Indo-West Pacific: Mauritius to Indonesia; Persian Gulf and Red Sea

Habitat: This benthic species living in coastal waters, estuaries, and freshwater lakes at a depth range of 16–26 m. It is amphidromous.

Maximum length: 25.0 cm; disc length, 140 cm

Food and feeding: It feeds mainly on bottom-living invertebrates.

Uses: Minor commercial fisheries exist for this species.

Pharmaceutical and nutraceutical compounds and activities

Antibacterial activity: Kalidasan et al. (2014) reported on the antibacterial activity of the chloroform extracts of the spine of this species. While maximum activity was with *S. aureus* (inhibition zone dm, 14 mm), minimum activity was with *E. coli* (1 mm) (Kalidasan et al., 2014).

Anticoagulant activity: The crude extracts from the spine of this species showed an anticoagulant activity of 91.50 USP units/mg against the standard heparin saline in which the activity was with 108.10 USP units/mg (Kalidasan et al., 2014).

Hypanus dipterurus (Garman, 1880) (*=Dasyatis brevis*)

Source: Image courtesy of © Philippe Béarez.

Common name(s): Whiptail stingray

Global distribution: Subtropical eastern Pacific: California to Peru; Hawaii

Habitat: This marine, demersal species is found in bays, in seagrass beds, kelp beds, and near reefs on sand and mud bottoms; depth range 1–70 m.

Maximum length and weight: 187 cm; 46.3 kg

Food and feeding: It is known to dig in the sand to feed. Its diet consists of small fishes, crabs, clams, and other benthic invertebrates.

Uses: Minor commercial fisheries exist for this species. It is also used in public aquariums.

Pharmaceutical and nutraceutical compounds and activities

Nutraceutical effects: The liver oil yield of this species is rich in triglyceride, sterol esters, free sterols, polar lipids, and diacyl glyceryl ethers; significant levels of carotene and tocopherol (6.9 mg/100 g, 25.3 mg/100 g, respectively); and DHA and EPA (16%). These factors make this oil suitable for human and animal nutrition (Navarro-Garcia et al., 2004).

Hypanus sabinus (Lesueur, 1824) (=*Dasyatis sabina*)

Source: Image by Coughdrop12. https://creativecommons.org/licenses/by-sa/4.0/deed.en

Common name(s): Atlantic stingray

Global distribution: Subtropical Western Atlantic regions, namely, Chesapeake Bay to Florida (USA) and Gulf of Mexico

Habitat: This marine, demersal species has a depth range of only 2–25 m. Adults normally inhabit coastal waters, including estuaries and lagoons and ascend rivers.

Maximum length and weight: 61.0 cm; 4.9 kg

Food and feeding: It feeds on anemones, polychaete worms, small crustaceans, clams, and serpent stars.

Uses: This species has no fisheries due to its venomous caudal spines. It is, however, sold as aquarium fish. It is also used in biomedical and physiological research.

Pharmaceutical and nutraceutical compounds and activities

Antibacterial activity: The bacterial species associated with this species possess antibacterial activities. The bacterial genera such as *Pseudomonas*, *Psychrobacter*, and *Stenotrophomonas* showed activity against *E. coli* and *B. cereus* exhibited activity against *E. coli*, MRSA, MSSA, and vancomycin-resistant enterococci (VRE) (Ritchie et al., 2017).

Maculabatis gerrardi (Gray, 1851) (=*Himantura gerrardi*)

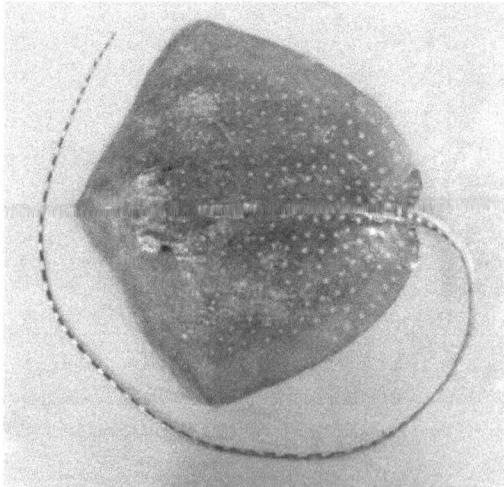

Source: Image by Hamid Badar Osmany. https://creativecommons.org/licenses/by/3.0/

Common name(s): Sharpnose stingray

Global distribution: Tropical Indo-West Pacific countries, namely, Taiwan and from Oman to Indonesia

Habitat: This marine species is demersal and it occurs in coastal waters at a depth of about 60 m. It is likely confined to the inner continental shelf, over sandy and mud bottoms. It has also been recorded from river mouths.

Maximum length: 200 cm

Food and feeding: It feeds mainly on bottom-living crustaceans such as shrimp, crabs, and small lobsters.

Uses: This species has commercial fisheries with game fish value.

Pharmaceutical and nutraceutical compounds and activities

Antimicrobial: Vennila et al. (2011) reported that the acidic extracts from the epidermal mucus of this species displayed peptide-derived antimicrobial action.

Pastinachus sephen (Forsskål, 1775) (=*Dasyatis sephen*)

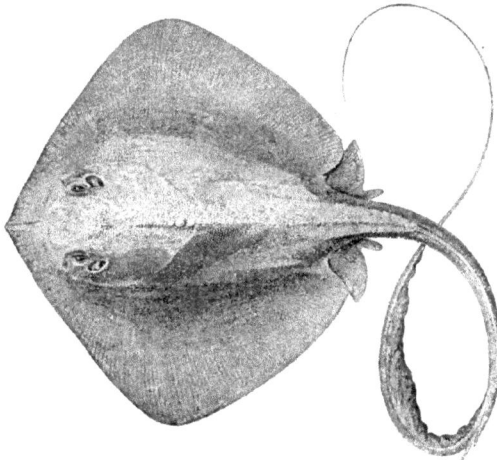

Source: Sir Francis Day. Public domain.

Common name(s): Cowtail stingray

Global distribution: Tropical NE Atlantic Ocean; Black Sea and Mediterranean Sea

Habitat: This species is found in the sandy or muddy habitats; and lagoons and reef flats at a depth range of 0–60 m. It also enters freshwater and brackish water as it is an amphidromous species.

Maximum length: 183 cm

Food and feeding: It feeds mainly on worms, shrimp, crab, and bony fishes.

Uses: Minor commercial fisheries exist for this species. It is also a commercially important aquarium fish.

Pharmaceutical and nutraceutical compounds and activities

Antibacterial and proteolytic activity: The acidic extracts from the epidermal mucus of this species showed antibacterial activity on *Salmonella typhi*, *Klebsiella pneumoniae*, *Streptococcus aureus*, *Escherichia coli*, and *Vibrio cholera*; and antifungal activity on *Aspergillus niger*, *Penicillium* sp., *T. mentagrophytes*, and *A. alternaria*. Further, the above extracts also showed proteolytic activity (Vennila et al., 2011).

2.2.8 BUTTERFLY RAYS (ORDER: MYLIOBATIFORMES; FAMILY: GYMNURIDAE)

Gymnura marmorata (Cooper, 1864)

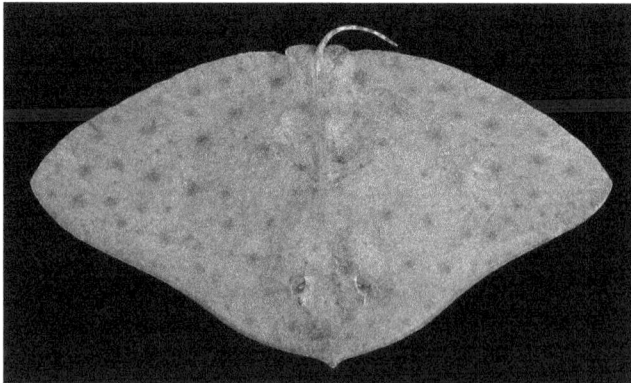

Source: Courtesy of Tomas Willems & Hans Hillewaert. https://en.wikipedia.org/wiki/Butterfly_ray#/media/File:Gymnura_micrura_.jpg
https://creativecommons.org/licenses/by-sa/3.0/

Common name(s): California butterfly ray

Global distribution: Subtropical eastern Pacific countries, namely, California (USA) to Peru

Habitat: This marine, demersal species inhabits shallow bays and beaches at a depth range of 1–94 m.

Maximum length: 100.0 cm

Food and feeding: It feeds mainly on crustaceans and small fishes.

Uses: This species has only minor commercial fisheries as it is sold mainly dried. It is also caught by recreational fishermen.

Pharmaceutical and nutraceutical compounds and activities

Nutraceutical effects: The liver oil yield of this species is rich in triglyceride, sterol esters, free sterols, polar lipids, and diacyl glyceryl ethers; significant levels of carotene and tocopherol (1.8 mg/100 g, 2.38 mg/100 g, respectively); and DHA and EPA (16%). These factors make this oil suitable for human and animal nutrition (Navarro-Garcia et al., 2004).

2.2.9 EAGLE RAYS (ORDER: MYLIOBATIFORMES; FAMILY: MYLIOBATIDAE)

Mobula hypostoma (Bancroft, 1831)

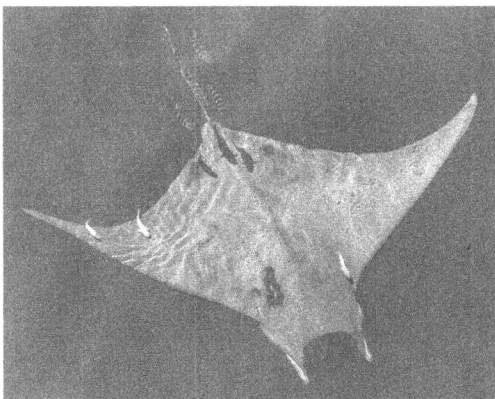

Common name(s): Lesser devil ray

Global distribution: Tropical Western Atlantic regions, namely, Brazil and Argentina; New Jersey (USA) to Santos; and St. Paul's Rocks of Eastern Atlantic

Habitat: This marine, pelagic-neritic species often occurs in shallow coastal waters and its depth range has been reported as 30–4000 m.

Maximum length: 120 cm

Food and feeding: It is a zooplankton feeder. However, crustaceans like shrimps and small fishes have also been reported from its stomach contents.

Uses: Minor commercial fisheries exist for this species. The dried gill plates of these rays can sell for hundreds of USD/kg and are traded globally for use in a Chinese medicinal tonic (http://www.sharkadvocates.org/pdf/facts/cites_devil_ray_fact_sheet.pdf).

Pharmaceutical and nutraceutical compounds and activities

Antibacterial activities: The bacterial species associated with this species possess antibacterial activities. Its *Vibrio* isolates showed activities against *B. subtilis,* VRE and MRSA; its *Pseudoalteromonas* isolates against *B. subtilis* and *Vibrio shilonii*; and its *Alteromonas* against *B. subtilis* only (Ritchie et al., 2017).

2.2.10 RIVER STINGRAYS (ORDER: MYLIOBATIFORMES FAMILY: POTAMOTRYGONIDAE)

Potamotrygon henlei (Castelnau, 1855)

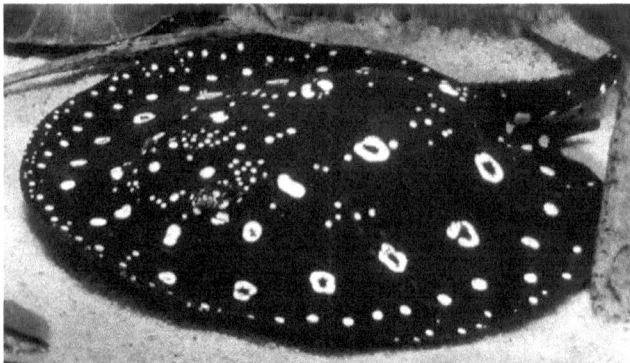

Source: Image by Christine Schmidt. https://creativecommons.org/licenses/by/2.0/

Common name(s): Bigtooth river stingray

Global distribution: Tocantins and Araguaia rivers in South America

Habitat: It is a freshwater, benthopelagic species.

Maximum length: 45.0 cm

Food and feeding: Feeds on fish and aquatic invertebrates, including worms and crustaceans.

Uses: It has no fisheries value. However, juveniles have aquarium values.

Pharmaceutical and nutraceutical compounds and activities

Antimicrobial activity: The purified fraction of protein derived from this species showed potent antimicrobial activity on *Micrococcus luteus* and weakest activity against *Escherichia coli* with MIC values of 4 and 12 µM, respectively. The protein also showed activity against the fungus species *Candida tropicalis* (Conceição et al., 2012).

Potamotrygon motoro (Müller & Henle, 1841)

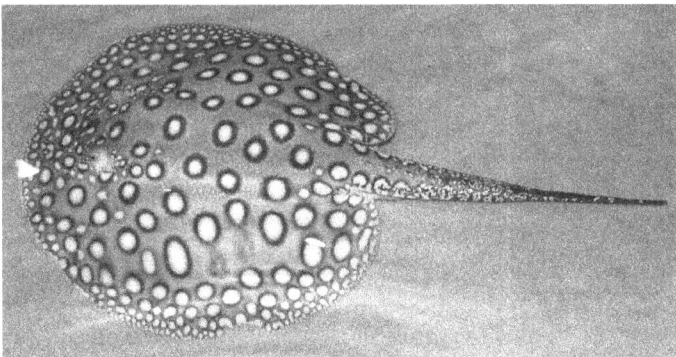

Source: Image by Jim Capaldi from Springfield, USA. https://creativecommons.org/licenses/by/2.0/

Common name(s): Peacock-eye stingray, ocellate river stingray or black river stingray

Global distribution: Throughout South American river systems

Habitat: It is a freshwater, benthopelagic species.

Maximum length and weight: 50.0 cm; 34.2 kg

Food and feeding: It feeds chiefly on annelids, mollusks, crustaceans, insect, and fish.

Uses: This species has only minor commercial fisheries with aquarium value.

Pharmaceutical and nutraceutical compounds and activities

Enzyme activity: The enzyme hyaluronidase derived from the venom of this species has medicinal values and it is used to increase drug diffusion and reverse the effects of hyaluronic acid filler injections (Cavallini et al., 2013; Magalhaes et al., 2008).

2.2.11   COWNOSE RAY (ORDER: MYLIOBATIFORMES; FAMILY: MYLIOBATIDAE)

Rhinoptera bonasus (Mitchill, 1815)

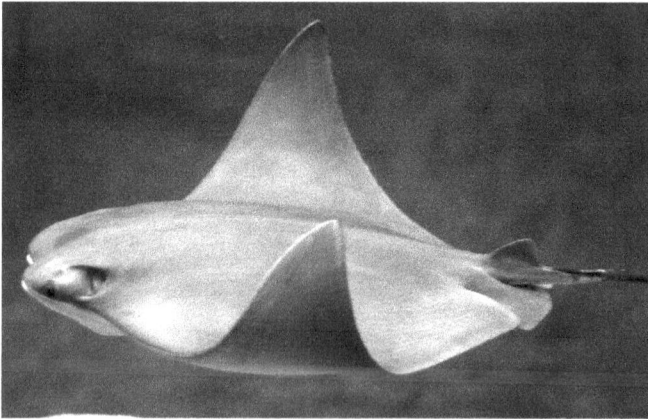

Source: Image by Todd Poling from Orient, USA. https://creativecommons.org/licenses/by/2.0/deed.en

Common name(s): Cownose ray

Global distribution: Tropical eastern Atlantic: Guinea, Mauritania and Senegal; western Atlantic: New England to Florida; and Gulf of Mexico, Brazil, Venezuela, and Trinidad

Habitat: This pelagic species is found marine and brackish habitats at a depth range of 0–22 m. It is oceanodromous and gregarious making long migrations.

Maximum length and weight: 213 cm; 953 g

Food and feeding: It feeds mainly on bottom-dwelling crabs, lobsters, mollusks, and bony fish.

Uses: This species has minor commercial fisheries with aquarium value.

Pharmaceutical and nutraceutical compounds and activities

Antibiotic activity: Bacterial strains isolated from the epidermal mucus of this species exhibited antibiotic activity on pathogenic bacterial strains including *Bacillus cereus*, *Bacillus subtilis*, *Staphylococcus aureus* (MSSA and MRSA), *Staphylococcus epidermidis*, *Micrococcus* sp., *Enterococcus faecalis*, *Listeria monocytogenes*, *Shigella boydii*, *Shigella sonnei*, *Shigella flexneri*, *Escherichia coli*, *Pseudomonas aeruginosa*, *Salmonella enterica*, *Acinetobacter baumannii*, and *Acinetobacter calcoaceticus* (Luer, 2012, 2013).

Ritchie et al. (2017) reported that several bacterial strains isolated from this species showed broadest spectra of antibacterial activity. Among these strains, *Bacillus* sp. (803E6) exhibited significant activity against MRSA, MSSA, VRE, and *B. subtilis* with inhibition zone values of 7.5, 8.5, 4.5, and 10 mm, respectively; *Halomonas* sp. (803D5) against MRSA (>10 mm); *Shewanella* sp. (806B10), *Alteromonas* sp. (806B11) against MSSA (>10 mm), and *Vibrio* sp. against VRE.

2.2.12 BAMBOO SHARKS (ORDER: ORECTOLOBIFORMES; FAMILY: HEMISCYLLIIDAE)

Chiloscyllium punctatum Müller & Henle, 1838

Source: Image by Christian Leonard Quale. https://creativecommons.org/licenses/by-sa/2.0/deed.en

Common name(s): Brownbanded bamboo shark

Global distribution: Tropical; Indo-West Pacific: Andaman Island and India east to the Philippines; north to Japan and south to Australia

Habitat: It is a common inshore bottom shark found on coral reefs, often in tide pools; depth range 0–85 m.

Maximum length: 132 cm

Food and feeding: It probably feeds on bottom invertebrates and small fish.

Uses: Commercial fisheries exist for this species. It is also a commercially important aquarium fish.

Pharmaceutical and nutraceutical compounds and activities

ASC and PSC have been isolated from the cartilage of this species. Both collagens had protein (89.8% and 89.9%, respectively) as a major constituent with the trace amount of ash and fat. Further, these collagens have significant amounts of amino acids such as alanine, proline, and hydroxyproline. Based on these factors, this species and its collagens may have biotechnological applications (Kittiphattanabawon et al., 2010).

2.2.13   SKATES (ORDER: RAJIFORMES; FAMILY: RAJIDAE)

Leucoraja erinacea (Mitchill, 1825)

Common name(s): Little skate

Global distribution: Temperate region of western Atlantic: Gulf of St. Lawrence; Nova Scotia (Canada) to North Carolina (USA)

Habitat: This benthic species lives in the gravely or sandy bottoms at a depth range of 0–329 m. It is most active during daytime.

Maximum length: 54.0 cm

Food and feeding: Its food organisms include polychaetes, crabs, shrimps, squids, sea squirts, and bony fishes

Uses: It is not a commercial species. However, it is used often to bait traps, especially lobster and eel traps.

Pharmaceutical and nutraceutical compounds and activities

The cells of immune tissue of this species displayed phagocytic activity and induction of apoptosis. Further, its lymphomyeloid tissue has the potential to serve as therapeutic agents (Luer and Walsh, 2018).

Okamejei kenojei **(Müller & Henle, 1841)** (*=Raja kenojei; Raja porosa*)

Source: OpenCage.info. https://creativecommons.org/licenses/by-sa/2.5/

Common name(s): Ocellate spot skate

Global distribution: Temperate; Northwest Pacific: Japan, Korea, and Taiwan; off Japan and in China Seas south to about Xiamen

Habitat: This benthic species is mostly restricted to with a depth range of 20–120 m. It is oceanodromous.

Maximum length and weight: 57.0 cm; 750 g

Food and feeding: Feeds on invertebrates and bony fish.

Uses: Commercial fisheries exist for this species.

Pharmaceutical and nutraceutical compounds and activities

ACE-inhibitory activity: The peptides derived from the skin of this species showed ACE-inhibitory activity with an IC_{50} value of 3.1 μM (Pangestuti and Kim, 2017).

Ngo et al. (2014) reported that the two peptides with the amino acid sequences of MVGSAPGVL and LGPLGHQ purified from the skin gelatin of this species showed ACE-inhibitory activity with IC_{50} values of 4.2 and 3.1 μM, respectively. These purified peptides offer vast scope for use in the food industry as functional ingredients with potent benefits in reducing the risk of cardiovascular diseases.

Anti-Alzheimer's and neuroprotective activity: The purified peptide (with the amino acid sequence of Gln–Gly–Try–Arg–Pro–Leu–Arg–Gly–Pro–Glu–Phe–Leu) derived from the skin neutrase hydrolysate of this species showed strong secretase inhibitory activity with an IC_{50} value of 24.3 μM. Further, the synthesized tetrapeptide with the amino acid sequence of Pro–Glu–Phe–Leu had also the highest secretase inhibitory activity. The result of this study suggests that the secretase inhibitory peptide derived from this skate skin may be beneficial as antidementia compounds in functional foods or as pharmaceuticals (Lee et al., 2015).

Antioxidant activity: Three hexapeptides were isolated from the protein hydrolysates of the cartilage of this species and their amino acid sequences were identified as Phe–Ile–Met–Gly–Pro–Tyr (I), Gly–Pro–Ala–Gly–Asp–Tyr (II), and Ile–Val–Ala–Gly–Pro–Gln (III). All these three peptides exhibited significant scavenging activities on DPPH (EC50, 3.6, 6.0, and 6.7 μM, respectively); HO (EC50, 4.2, 6.8, and 8.6 μM, respectively), O_2 (EC50, 2.2, 2.9, and 3.1 μM, respectively), and ABTS (EC50, 1.4, 1.3, and 2.2 μM, respectively). Further, the peptide III was also found to be effective against lipid peroxidation. These findings suggest that the said peptides have excellent antioxidant properties for use as food additives and functional foods (Pan et al., 2016a).

Anticancer activity: The protein hydrolysate of the cartilage of this skate yielded a hexapeptide with amino acid sequence of Phe–Ile–Met–Gly–Pro–Tyr. This peptide exhibited significant anticancer activity on HeLa cells with an IC_{50} value of 4.8 mg/mL. Further, this peptide has been reported to induce apoptosis (Pan et al., 2016b).

Raja clavata **Linnaeus, 1758**

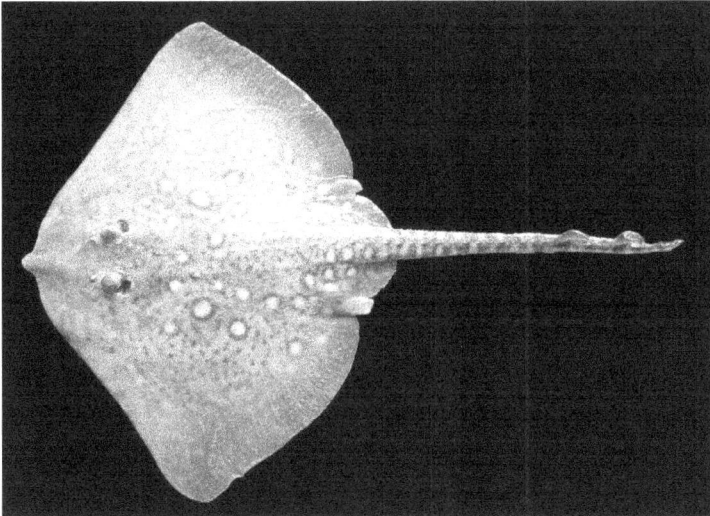

Source: Image by Hans Hillewaert. https://creativecommons.org/licenses/by-sa/4.0/

Common name(s): Thornback ray

Global distribution: This is found distributed in the subtropical regions of Mediterranean and northeastern Atlantic

Habitat: This benthic species inhabits mud, sand, and gravel bottoms of the shelf and upper slope waters with a depth range of 5–1020 m.

Maximum length and weight: 105 cm; 18 kg

Food and feeding: It is a nocturnal species feeding on all kinds of bottom animals, preferably crustaceans and fishes.

Uses: This species has commercial fisheries with game fish value.

Pharmaceutical and nutraceutical compounds and activities

ACE-inhibitory and antioxidant activity: The hydrolysates derived through the protease from *Bacillus subtilis* A26 (TRGH-A26) and neutrase from *Bacillus amyloliquefaciens* (TRGH-Neutrase) exhibited ACE-inhibitory activity with IC_{50} values of 0.94 and 2.07 g/L. These hydrolysates also showed DPPH radical scavenging activity with IC_{50} values of 1.98 and 21.2 g/L, respectively. Further, the peptides purified from the above hydrolysates showed the amino acid sequences of APGAP (TRGH-A26) and GIPGAP (TRGH-Neutrase) and exhibited ACE activity with IC_{50} values of 170 and 27.9 µM, respectively (Lassoued et al., 2015).

Raja eglanteria Bosc, 1800

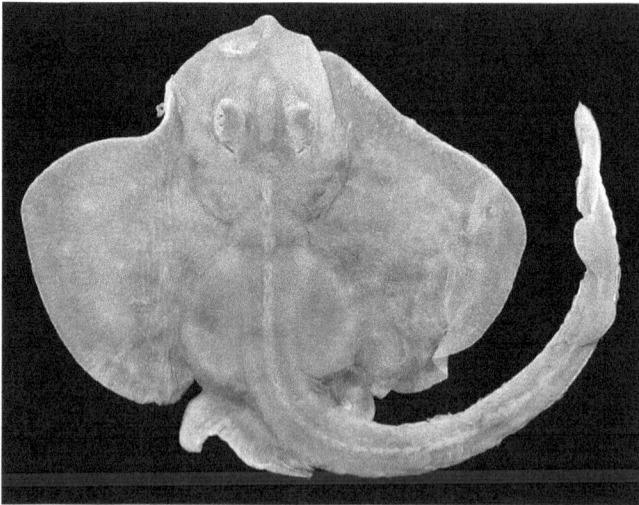

Source: Smithsonian Environmental Research Center. https://creativecommons.org/licenses/by/2.0/deed.en

Common name(s): Clearnose skate

Global distribution: Subtropical western north Atlantic regions from Massachusetts to Florida; Gulf of Mexico

Habitat: This benthic species commonly inhabits inshore areas with a depth range of 0–330 m. It is also seen in estuaries and bays. It is also oceanodromous moving offshore areas during the colder months.

Maximum length: 84.0 cm

Food and feeding: It feeds mainly on decapod crustaceans, bivalves, polychaetes, squids, and fishes.

Uses: Minor commercial fisheries exist for this species. Occasionally, it is used as bait.

Pharmaceutical and nutraceutical compounds and activities

Antibacterial activity: Seven bacterial isolates derived from the epidermal mucus of this species showed antibacterial activity against *E. coli* with inhibition zone values ranging from 2 to 4 mm. One isolate was also active against MRSA with zone of inhibition value of 4 mm (Luer, 2014). Luer (2013) also reported that its fresh mucus possessed weak antimicrobial activity, and mucus-containing symbiotic bacteria showed significant activity against MRSA and VRE.

2.2.14   GULPER SHARKS (ORDER: SQUALIFORMES; FAMILY: CENTROPHORIDAE)

Centrophorus grunulosus (Bloch & Schneider, 1801)

Source: Image by D Ross Robertson. Public domaine.

Common name(s): Gulper shark

Global distribution: Temperate; Western North Atlantic: Gulf of Mexico; Eastern Atlantic: Portugal, France, Madeira to Mediterranean; Ivory Coast, Senegal, Nigeria; Cameroon to Zaire; Western Indian Ocean: Aldabra Island; western Pacific: Japan

Habitat: This bathydemersal, fully marine species can be found offshore with a depth range of 50–1200 m. It is a highly migratory species and has schooling habits.

Maximum length: 150 cm

Food and feeding: Its major food items include bony fish, cephalopods, and crustaceans.

Uses: Minor commercial fisheries exist for this species. It can be smoked and/or dried/salted for human consumption and is also processed for fish-meal and liver oil.

Pharmaceutical and nutraceutical compounds and activities

Nutraceutical properties: This species is known for its liver oil which has high squalene content with several therapeutic applications (FAO, http://www.fao.org/fishery/species/2835/en).

Centrophorus squamosus (Bonnaterre, 1788)

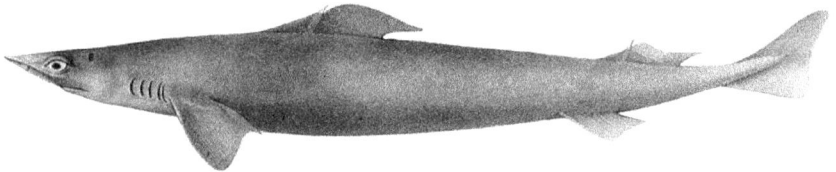

Source: Image by Müller & Henle. Public domain.

Common name(s): Leafscale gulper shark

Global distribution: Eastern Atlantic: from Iceland south to Cape of Good Hope; Western Indian Ocean: Aldabra Islands; western Pacific: Japan, Honshū, the Philippines, Australia, and New Zealand

Habitat: This marine, bathydemersal species has a depth range of 145–2400 m. This deepwater species is found on or near the bottom of continental slopes; also found pelagically in the upper 1250 m of water 4000 m deep.

Maximum length: 164 cm

Food and feeding: It feeds on cephalopod mollusks and bony fishes.

Uses: This species has only minor commercial fisheries and it is sold dried-salted. It is also used for fishmeal production.

Pharmaceutical and nutraceutical compounds and activities

Antimicrobial activity: The crude homogenate (CH) in liver, kidney, spleen and pancreas; and gray fat (GF) fraction of liver from aminosterol extraction of this species was tested antimicrobial activity against *Staphylococcus aureus*, *Escherichia coli*, *Enterobacter cloacae*, *Streptococcus pyogenes*, and *Candida albicans*. Liver GF samples from this shark showed strong activity, against all bacteria and yeast. The liver samples from this shark showed strong activity against *E. coli* and *C. albicans* and moderate activity against *S. aureus*. Further, the water fraction from lipid extraction in tissues from this species also showed antimicrobial activity (Remme et al., 2005).

Nutraceutical properties

Liver oil: The 1-*O*-alkylglycerol ether lipids of the liver oil of this species has been reported to display chemoprotective properties against reactive oxygen species; and antibacterial, antifungal, anticancer, and anti-inflammatory activities (Bordier et al., 1996; Zhang et al., 2013).

Fatty acids: This species is known for its nutritionally important fatty acids namely saturated, monounsaturated, and polyunsaturated fatty acids which are present at 33.8%, 34.6%, and 36.1%, respectively, and DHA at 12.9% (Remme, 2005).

Squalene content: Further, its liver contains the highest squalene content (70.6%). The values of squalene content in the tissues from the different organs of this species are given below.

Squalene content (%) in tissues from the organs of *C. grunulosus.*

Organ	%
Liver	70.6
Stomach	52.5
Pancreas	45.6
Heart	43.0
Spleen	30.1
Kidney	5.3

Source: Remme (2005).

2.2.15　BRAMBLE SHARKS (ORDER: SQUALIFORMES; FAMILY: ECHINORHINIDAE)

Echinorhinus brucus (Bonnaterre, 1788)

Source: Image by Arthur Bartholemew. Public domain.

Common name(s): Bramble shark

Global distribution: Western Atlantic: Virginia and Massachusetts (USA); Argentina and Venezuela; Eastern Atlantic: Morocco to Cape of Good Hope, North Sea to Mediterranean, and South Africa; Indian Ocean: India, Mozambique, and South Africa; western Pacific: Japan, southern Australia, New Zealand; and eastern Pacific

Habitat: It is a benthic species living on the continental and insular shelves and upper slopes at depth range of 10–900 m.

Maximum length: 310 cm

Food and feeding: Its major food items include crabs, smaller sharks, and teleost fishes.

Uses: Minor commercial fisheries exist for this species which is a game fish also.

Pharmaceutical and nutraceutical compounds and activities

Anti-inflammatory and anti-ulcer activities: The oil of this species showed significant proportion of n-3 PUFAs and the percentages of EPA and DHA were 16% and 18%, respectively. Its liver oil had a very favorable n3:n6 ratio of 4.7. Oral administration of the liver oil at 1 g/kg concentration showed anti-inflammatory activity by significantly attenuating the formalin-induced paw edema in experimental rats with 43.5%, 48.4%, and 47.8%

after 1st, 2nd, and 3rd hour, respectively. It also exerted potent anti-ulcer effects against acid–ethanol mixture mediated lesion formation in the rat gastric mucosa (Vishnu et al., 2015).

Cytotoxicity: The proteoglycans isolated from the cartilage of this species showed significant cytotoxicity against MCF-7 cell lines in a dose-dependent manner with an IC_{50} value of 25 µg/mL and apoptosis. The cell cytotoxicity was 73% at a concentration of 100 µg/mL (Ajeeshkumar et al., 2017). Vishnu et al. (2016) also reported that the oil of this species showed significant in-vitro cytotoxic effect on the human neuroblastoma cell line (SHSY-5Y) with IC_{50} values ranging from 35 to 45 ng.

Nutraceutical effects: The nutritional composition of the liver oil of this species showed essential fatty acids such as stearic acid (8%), palmitic acid (15%), oleic acid (12%), EPA (16%), and DHA (18%); vitamins A, D, and K at 17.1, 15.0, and 11.5 mg/100 g of oil, respectively; and squalene (38.5%). These findings show that this oil could be a potential source of functional foods to combat malnutrition (Vishnu et al., 2016).

2.2.16 LANTERN SHARKS (ORDER: SQUALIFORMES; FAMILY: ETMOPTERIDAE)

Centrocyllium fabricii (Reinhardt, 1825)

Source: Image by I, Tambja. https://creativecommons.org/licenses/by-sa/3.0/

Common name(s): Black dogfish

Global distribution: Atlantic: Iceland and Greenland to Virginia; southwestern and west Africa; and Argentina

Habitat: This offshore species is found in the continental shelf and continental slope at a depth range of 180–250 m. It makes seasonal migrations to shallow water during spring and winter.

Maximum length and weight: 75 cm

Food and feeding: It is an opportunistic predator and scavenger and is mainly feeding on crustaceans, cephalopods, and bony fishes.

Uses: It has minor commercial fisheries and it is mainly utilized for fish-meal production.

Pharmaceutical and nutraceutical compounds and activities

Antimicrobial activity: The CH in liver, kidney, spleen, and pancreas; and GF fraction of liver from aminosterol extraction of this species were tested antimicrobial activity against *Staphylococcus aureus*, *Escherichia coli*, *Enterobacter cloacae*, *Streptococcus pyogenes*, and *Candida albicans*. Among the different tissues, its liver exhibited significant antibacterial activity on all pathogens excepting *E. coli* which had weak activity; and the spleen and pancreas samples from this fish had strong activity against all bacteria and yeast. Further, antimicrobial activity was also evident with the water fraction from lipid extraction in tissues from this species (Remme et al., 2005).

Nutraceutical effects: The nutritionally important fatty acids of this species showed the presence of saturated, monounsaturated, and polyunsaturated fatty acids at 33.96%, 38.18%, and 33.76%, respectively. Further, the squalene content of this species was 17.2%. Among the different tissues, the liver showed the highest squalene content (48.8%). The values of squalene content in the tissues from the different organs of this species are given below.

Squalene content (%) in tissues from the different organs of black dogfish.

Organ	%
Heart	5.2
Liver	48.8
Stomach	35.0
Spleen	12.2
Pancreas	0.4
Kidney	1.3

Source: Remme (2005).

2.2.17 SLEEPER SHARKS (ORDER: SQUALIFORMES; FAMILY: SOMNIOSIDAE)

Centroscymnus coelolepis **Barbosa, du, Bocage, de, Brito & Capello, 1864**

Source: From plate 4 of Oceanic Ichthyology by G. Brown Goode and Tarleton H. Bean, published 1896. Public domain.

Common name(s): Portuguese dogfish

Global distribution: Western Atlantic: Grand Banks to Delaware, USA; Cuba; Eastern Atlantic: Iceland south along Atlantic slope to the southwestern Cape coast of South Africa; also western Mediterranean. Western Pacific: off Japan, New Zealand, and Australia; Western Indian Ocean: Seychelles

Habitat: This marine, bathydemersal species is found at depths of 128–3700 m in the continental slopes and abyssal plains.

Maximum length: 120 cm

Food and feeding: It feeds mainly on fish (including sharks) and cephalopods; also gastropods and cetacean meat.

Uses: Minor commercial fisheries exist for this species. It is dried and salted for human consumption. It is also utilized as fishmeal and as a source of squalene, a bioactive compound.

Pharmaceutical and nutraceutical compounds and activities

Antiproliferative activity: The hydrolysate of this species containing 96.8% peptide-induced growth inhibition on the human breast cancer cell line, MDA-MB-231 (Picot et al., 2006).

Antimicrobial activity: The CH derived from the liver, kidney, spleen, and pancreas; and fat fraction of liver from aminosterol extraction of this species were tested antimicrobial activity against *Staphylococcus aureus*, *Escherichia coli*, *Enteriobacter cloacae*, *Streptococcus pyogenes*, and *Candida albicans*. The samples from this dogfish showed strong activity against *S. aureus* and *S. pyogenes*, weak activity against *Enterobacter cloacae*, and no activity against *E. coli* and *C. albicans*. While there was no antimicrobial activity in the kidney samples, the spleen exhibited moderate activity with most of the pathogens. Among the water fractions, only the stomach and heat samples showed activity against *S. pyogenes* and *E. cloacae* (Remme et al., 2005).

2.2.18   DOGFISH SHARKS (ORDER: SQUALIFORMES; FAMILY: SQUALIDAE)

Squalus acanthias Linnaeus, 1758

Source: Image by Doug Costa, NOAA/SBNMS. Public domain.

Common name(s): Picked dogfish

Global distribution: Temperate Northern and Southern Hemispheres

Habitat: Habitat: It is found in the inshore and offshore waters a depth range of 0–1460 m. It rarely enters brackish water as it prefers seawater. It is highly migratory forming large schools.

Maximum length and weight: 95.0 cm; 9.1 kg

Food and feeding: Its food items include comb jellyfish, shrimps, crabs, squids, and sea cucumbers; mackerel and herring; and a variety of benthic fishes.

Uses: This species has commercial fisheries. It is largely consumed in several countries such as USA, Canada, Europe, and New Zealand. Its fins and tails are used in Chinese cuisine. It is also utilized as liver oil and fertilizer.

Pharmaceutical and nutraceutical compounds and activities

Squalamine: This shark contains the squalamine, a steroid which has been reported to be a potent antibacterial, antiviral, anticancer, and immunomodulatory activities (Khora, 2013). The various bioactivities shown by this compound are detailed below.

Anticancer activity: Patients with advanced lung and ovarian cancers when treated with squalamine at 192 mg/m^2/day continuously for 120 h experienced hepatotoxicity. These findings suggest that squalamine may be of great use in chemotherapy (Bhargava et al., 2001). Li et al. (2002) reported that squalamine has the ability to stop MAP kinase and associated cell proliferation in vascular endothelial cells.

Antimicrobial activity: Squalamine has been reported to exhibit antimicrobial activity on both Gram-negative and Gram-positive bacteria and the values recorded are given below.

Antimicrobial activity (MIC, µg/mL)

Proteus vulgaris	4–8
Candida albicans	4–8
Paramecium caudatum	4–8
Escherichia coli	1–2
Pseudomonas aeruginosa	4–8
Staphylococcus aureus	1–2
Streptococcus faecalis	1–2

Source: Moore et al. (1993).

Antiviral activity (Zasloff et al., 2011)

Dengue virus Den V2 inhibitory activity: The squalamine of this species exhibited broad-spectrum activity on human microvascular endothelial cells inoculated with dengue virus Den V2. At a squalamine concentration of 100 g/mL, complete inhibition was noted. The percentage inhibition values recorded at different concentrations of squalamine are given below.

Squalamine (µg/mL)	% Inhibition
0	38
10	36
20	26
40	15
60	9
100	0

Source: Zasloff et al. (2011).

Human hepatitis B virus inhibitory activity: Squalamine has been reported to inhibit human hepatitis B virus replication in human primary hepatocytes. The values of percentage inhibition and HBV replication recorded at different concentrations of squalamine are given below.

Squalamine (µg/mL)	% HBV replication	% Inhibition
2	86	14
6	46	54
20	16	84

Source: Zasloff et al. (2011).

2.3 JAWLESS FISHES (PHYLUM, CHORDATA; SUBPHYLUM, VERTEBRATA; CLASS: AGNATHA)

2.3.1 *HAGFISH (ORDER: MYXINIFORMES; FAMILY: MYXINIDAE)*

Myxine glutinosa Linnaeus, 1758

Source: Image by Arnstein Rønning, https://creativecommons.org/licenses/by-sa/3.0/deed.en

Common name(s): Atlantic hagfish

Global distribution: Boreal; North Atlantic: Murmansk to the Mediterranean Sea; Greenland to USA

Habitat: This marine, benthopelagic, and nonmigratory species has a depth range of 20–1200 m. It is found on muddy bottoms where they hide in the mud.

Maximum length: 95.0 cm

Food and feeding: The major food items of this nocturnal fish are dead fishes.

Uses: It has no fisheries value.

Pharmaceutical and nutraceutical compounds and activities

Antimicrobial activity: An antimicrobial peptide, myxinidin, from the acidic epidermal mucus extract of this species yielded a peptide called myxinidin which exhibited antimicrobial activity on several species of bacteria and yeast. Pathogens such as *Escherichia coli* D31, *Salmonella enterica serovar typhimurium* C610, *Aeromonas salmonicida* A449, *and Listonella anguillarum* 02-11 registered MBC values ranging from 1.0 to 2.5 µg/mL, and *Staphylococcus epidermis* C621 and *Candida albicans* C627 had an MBC of 10.0 µg/mL (Subramanian et al., 2009).

Unidentified Hagfish (Slime Eel)

Aphrodisiac properties: The flesh of this fish broiled in sesame oil and sprinkled with salt was an appetizer among older Korean men. It was believed to give them energy like Viagra (http://www.nbcnews.com/id/19336602/ns/health-sexual_health/t/despite-ick-factor-slime-eel-has-sex-appeal/#.XjwXZTFKhPY).

2.3.2 LAMPREYS (ORDER: PETROMYZONTIFORMES; FAMILY: PETROMYZONTIDAE)

Entosphenus tridentatus (Richardson, 1836)

Source: Image by Dave Herasimtschuk, US Fish & Wildlife Service. Public domain.

Common name(s): Pacific Lamprey

Global distribution: Temperate areas of Japan, USA (California, Oregon, Washington, and Idaho) and British Columbia

Habitat: Marine; freshwater; brackish; this demersal, anadromous species has been reported to live in marine, freshwater, and brackish water areas at depths of 0–1508 m.

Maximum length and weight: 76.0 cm; 500 g

Food and feeding: It is a parasitic fish. Adults prey on fishes and sperm whales.

Uses: Subsistence fisheries exist for this species. It is occasionally used as bait.

Health benefits: A lady patient with a rare neurological condition (like the diabetic neuropathy with loss of feeling in arms and legs) became normal with the concept of the physiology of this lamprey which has been explained as Pacific lamprey fish keep swimming even if their spine is severed because they don't receive pain signals. Doctors made a connection and placed a circuit board implant in the back of the aforesaid patient to block pain like the lamprey (YakTriNews.com)

Lampetra fluviatilis **(Linnaeus, 1758)** (=*Petromyzon fluviatilis*)

Source: Image by Tiit Hunt. https://creativecommons.org/licenses/by-sa/3.0/

Common name(s): River lamprey

Global distribution: Norway to France; Baltic and Mediterranean

Habitat: Marine; freshwater; brackish; this demersal and anadromous species lives in hard bottoms of marine, freshwater, and brackish water areas at a depth of about 10 m. In freshwaters, it is found in rivers, brooks, and lakes. Some populations are permanent freshwater residents. It is known to attach to fish like herring and cod.

Maximum length and weight: 50.0 cm; 150.0 g

Food and feeding: Adults of this species are parasitic preying on fishes by sucking their blood and consuming their flesh.

Uses: This species has only minor commercial fisheries and it is often utilized as bait.

Nutraceutical properties: The oil of this species has been reported contains high iodine level than cod-liver oil. Costwise, it is eight times less than cod-liver oil. Further, this oil is well received by the digestive organs. It promotes nutrition even with a greater effect than that of cod-liver oil (The Druggists Circular and Chemical Gazette, 1878).

Unidentified lamprey

Brain therapy: Lampreys have been reported to help boost brain therapies. The molecules derived from the lampreys' immune system called "variable lymphocyte receptors" have been reported to carry drugs to the brain, boosting the effectiveness of treatments for brain cancer, brain trauma, or stroke (Cohut, 2019).

KEYWORDS

- species profile
- global distribution
- biology
- ecology
- pharmaceutical compounds
- nutraceutical compounds
- bioactivities

CHAPTER 3

Pharmaceuticals and Nutraceuticals from Fish Wastes and Their Activities

ABSTRACT

This chapter deals with the chemical composition of fish wastes; by-products derived from fish wastes; and pharmaceutical and nutraceutical applications of fish collagen and gelatin, protein hydrolysates, fish oil, enzymes, shark liver oil, shark cartilage, and so on.

In fish, the discards (wastes) such as fins, heads, skin, and viscera constitute about 50%. It is possible to convert these wastes into therapeutically important proteins, protein hydrolysates, lipids, vitamins, and minerals. Among these components, proteins and fish oils with their rich polyunsaturated fatty acids possess commercial value. Fish processing industries may convert the said fish wastes into valuable fisheries by-products, thereby aquatic environments where they are normally discharged may be safeguarded from pollution problems such as eutrophication and associated nutrient enrichment (Santhanam, 1990).

Composition of fish.

Component	Average weight (%)
Fillet, skinned	36
Fish wastes	64
Skin	3
Head	21
Backbone	14
Fins and lungs	10
Gut	7
Liver	5
Roe	4
Total	100

Source: Ghaly et al. (2013).

3.1 CHEMICAL COMPOSITION OF FISH WASTES

It is estimated that the wastes of fish processing industries such as protein, lipid, and monounsaturated fatty acids such as oleic acid and palmitic acid constitute 58%, 19%, and 22%, respectively. These wastes may be profitably converted into valuable by-products such as proteins, amino acids, collagen, gelatin, oil, and enzymes. Further, fish frames for a rich source of muscle proteins.

Proximate composition and mineral contents of fish wastes.

Proximate composition	
Protein (%)	57.9
Fat (%)	19.1
Fiber (%)	1.2
Ash (%)	21.8
Minerals	
Calcium (%)	5.8
Phosphorous (%)	2
Potassium (%)	0.7
Sodium (%)	0.6
Magnesium (%)	0.2
Manganese (ppm)	6.0
Iron (ppm)	100
Copper (ppm)	1
Zinc (ppm)	62

Source: Ghaly et al. (2013).

Proximate composition (%) of discards (such as head, frame, and backbone) of Flathead, Salmon, Barramundi, and Silver warehou.

Fish*	Protein	Fat
Flathead	14.7	3.3
Salmon	14.8	18.9
Barramudi	1.8	16.9
Silver warehou	15.7	1.9

*Flathead (*Platycephalus fuscus*), Atlantic salmon (*Salmo salar*), Barramundi (*Lates calcarifer*), and Silver warehou (*Seriolella punctata*).

Source: Nurdiani et al. (2015).

Fatty acids of discards (such as head, frame, and backbone) of Salmon, Flathead, Silver warehou, and Barramundi.

Fatty acids (%)	Salmon	Flathead	Silver warchou	Barramundi
SFA	30.6	49.2	46.8	33.2
MUFA	3.4	7.5	7.7	3.8
PUFA	11.8	1.9	1.2	20.2

SFA, Saturated fatty acids; MUFA, mono-unsaturated fatty acids; PUA, polyunsaturated fatty acids.

Source: Nurdiani et al. (2015).

Proximate composition of freshwater fish discards (such as skin, fins, and scales).

Species	Carbohydrate (%)	Protein (%)	Fat (%)
Cyprinus carpio			
Skin	2.9	28.1	2.0
Scales	6.1	29.9	0.9
Fins	4.7	19.2	7.0
Hypophthalmichthys molitrix			
Skin	2.9	28.1	2.0
Scales	5.1	28.4	0.9
Fins	7.2	21.4	8.8

Source: Mahboob et al. (2014).

3.2 BY-PRODUCTS OF FISH WASTES

Amino acids: As fish discards consist of amino acids in well-balanced level and they offer vast scope in the manufacture of value-added pharmaceutical and nutraceutical products.

3.2.1 APPLICATIONS OF AMINO ACIDS DERIVED FROM FISH WASTES

Amino acids have wide applications such as in food additives, feed, and food supplements and in the pharmaceutical industries. It has been reported that amino acids such as arginine, histidine, glycine, and glutamate are presently utilized as an excipient in the development of drugs. Amino acids such as alanine, aspartate, arginine, and monosodium glutamate find applications in the food flavoring industries. Further, several amino acids derived from fish could be employed in the production of antibiotics (Ghaly et al., 2013).

Amino acid composition of fish discards (g/100 g crude protein).

Amino acid	Marine fish wastes	Freshwater fish wastes
Threonine	2.9	3.2
Serine	2.7	3.4
Tryptophan	0.8	1
Lysine	10.1	7.5
Histidine	5.2	2.7
Arginine	3.0	3.6
Aspartic acid	9.1	10.2
Glutamic acid	13.6	16.2
Proline	3.2	4.4
Methionine	6.9	3.2
Isoleucine	5.3	5.4
Leucine	9.2	9.6
Glycine	6.5	6.2
Alanine	8.6	9.3
Cystine	0.8	1
Valine	6.4	6
Tyrosine	1.8	2.4
Phenylalanine	4	5.0

Source: Ghaly et al. (2013).

3.2.2 PHARMACEUTICAL AND NUTRACEUTICAL IMPORTANCE OF AMINO ACIDS FROM FISH PROCESSING WASTES

i) **Amino acids of pharmaceutical importance:** cysteine glycine, glutamine, arginine, valine, leucine, alanine, isoleucine, histidine, proline, serine, and tyrosine

ii) Amino acids of nutraceutical importance

Glutamate	Flavor enhancer
Alanine	Flavor sweetener
Methionine	Feed supplement
Lysine	Feed supplement
Threonine	Feed supplement
Tryptophan	Feed supplement

Source: Ikeda (2003).

3.2.3   PHARMACEUTICAL APPLICATIONS OF AMINO ACIDS DERIVED FROM FISH PROCESSING WASTES

Approved protein-based pharmaceuticals containing amino acids.

Amino acids	Product	Drug substance
Histidine	Recombinate	Recombinate antihemophilic factor
Histidine	KogenateFS	Recombinate antihemophilic factor
Histidine	Synagis	Human monoclonal antibody
Histidine	Herceptin	Human monoclonal antibody
Histidine	BeneFIX	Coagulation factor IX
Arginine	TNKase	Human tissue plasminogen activator
Arginine	Activase	Human tissue plasminogen activator
Glycine	KogenateFS	Recombinant antihemophilic factor
Glycine	BeneFIX	Coagulation factor IX
Glycine	Nutropin	Human growth hormone
Glycine	Neumega	Interleukin-11
Glutamate	VariVax	Varicella virus vaccine live
Glutamate	Streptase	Streptokinase

Source: Arakawa et al. (2007).

Bioactive peptides: The proteins derived from the enzyme hydrolysis of fish muscle yield a number of peptides with bioactivities such as antihypertensive, antithrombotic, immune-modulatory, anti-oxidative, anticoagulant, and antiplatelet properties. The bioactive peptides obtained from the fish muscle have anticoagulant and antiplatelet properties (Cheung et al., 2015).

Fish peptide products, their source, and applications.

Product	Source	Applications
Gabolysat PC60®/Stabilium®/ Protizen®/Procalm®	Hydrolysate of fish protein	Anxiolytic
Seacure®	Hydrolysate of fish protein	Intestinal health
Nutripeptin®/Hydro MN peptide	Hydrolysate of fish protein	Postprandial blood glucose control
Katsuobushi oligopeptide	Hydrolysate of dried bonito	Antihypertensive
Fish gelatin	Hydrolysate of fish collagen and gelatin	Nutrient supplements

Source: Cheung et al. (2015).

Gelatin and collagen: The gelatin and collagen derived from the skin wastes have a number of biotechnological, pharmaceutical, and nutraceutical applications.

Fish oil: The oil derived from the fish wastes consists of two important polyunsaturated fatty acids, namely, eicosapentaenoic acid, and docosahexaenoic acid and forms an important component in human nutrition due to its health-related bioactivities.

Enzymes: The internal organs of the fish are a rich source of enzymes which include mainly several enzymes such as pepsin, trysin, chymotrypsin, and collagenase are obtained from the internal organs of fish.

Minerals: Fish bones mainly form the major source of minerals, such as calcium, phosphorous, and hydroxyapatite, and these minerals constitute about 70%.

3.3   BONY FISHES

3.3.1   FISH PROTEINS

The fish processing wastes yield protein by-products such as fish sauce and collagen and its derivative gelatin, and protein hydrolysates.

3.3.1.1   COLLAGEN AND GELATIN

High-quality collagens derived from marine fishes possess good absorption capacity and are therefore largely used in biomedical and biomaterial applications.

Fish wastes as sources of gelatin: The fish skin, scales, and bones could be the alternative sources of collagen which amounts to 36–54%. Though the sardine scales are the potential source of collagen, other marine fishes such as salmon, cod, sole, hake, and shark, and freshwater fishes like catla, rohu, and tilapia are also largely used in the extraction of collagen (Bernhardt et al., 2018).

Types of collagen: The collagen of fish is of three types, namely, type I present in the skin, bone, and tendons; type II collagen in the cartilage tissue; and type III in the skin of young fish. The functional properties of these collagens largely depend on their amino acid composition, particularly in the concentrations of imino acids (proline and hydroxyproline). Generally, the collagen of warm water fishes such as bigeye-tuna and tilapia possess a higher imino acid level that of cold-water fishes, such as cod, whiting, and halibut (Eastoe and Leach, 1977).

Gelatin: Fish gelatin is obtained from its fibrous protein collagen. The skin of snakeheads, tilapia, carps, and catfish; tuna head bone and fish swim bladder are the major sources of gelatin with high gel strength. Gelatin is derived by heating the collagen in acid, alkali, or enzyme. Ezymatic hydrolysis improves further degradation of gelatin into its hydrolysates. Hot water treatment may also be used to solubilize collagen of the skin and extract it as gelatin. The gelatin type depends on the fish type and production process. For example, the gelatin of cold-water fishes, such as pollock, haddock, cod, and salmon has poor content of imino acids. The gel strength and melting temperature of fish gelatins vary in their amino acid composition (Karim and Bhat, 2009).

Extraction and uses of gelatin: Though the gelatin is obtained from collagen through acid and alkali processes, the acidic treatment has been found to be the best especially for the collagens of fish skins and bone raw materials. The major processing steps in gelatin production involve (i) pretreatment of raw materials, (ii) hydrolysis of collagen to gelatin, and (iii) purification and drying. Gelatin obtained from freshwater and marine fishes serves as a biopolymer which is used as a food ingredient. Cold-water fish gelatin is used in human consumption as frozen or refrigerated products. Fish gelatin is also used as colorants in microencapsulation and as food flavors (Karim and Bhrat, 2008).

3.3.1.1.1   Processing Steps for the Production of Gelatin from Fish Skin and Bone (Shahidi et al., 2019)

Acid process: By this process, the fish skin is processed for the production of gelatin as follows:

Fish skin > degreasing > acid treatment > extraction > filtration > ion exchange > concentration > sterilization > drying > gelatin

Alkali process: By this process, the fish bone is processed for the production of gelatin as follows:

Fish bone > degreasing > demineralization > liming > neutralization > extraction > filtration > ion exchange > concentration > sterilization > drying > gelatin

Collagen content of marine bony fishes (values in %; wet wt basis)

Species	Source	Yield (%) (ww)	References
Acanthopagrus latus	Bone	40.1	Nagai and Suzuki (2000)
Aluterus monoceros	Skin	ASC 4.19	Ahmad et al. (2010)
Anchoviella sp.	White muscle	ASC 0.09	Mathew and Hassan (1996)
Anchoviella sp.	White muscle	ASC 0.09	Mathew and Hassan (1996)
Archosargus probatocephalus	Skin	ASC 2.6; PSC 29.3	Ogawa et al. (2003)
Brama australis	Skin	1.5	Sionkowska et al. (2015)
Caranx spp.	White muscle	ASC 0.50	Mathew and Hassan (1996)
Cynoglossus semifasciatus	White muscle	ASC 0.40	Mathew and Hassan (1996)
Cypselurus melanurus	Scales	ASC 0.72	Thuy et al. (2014)
Diodon holocanthus	Skin	ASC 4; PSC 19.3 (d.w.)	Huang et al. (2011)
Epinephelus lanceolatus	Skin	ASC 39.51; PSC 19.12	Hsieh et al. (2014)
Evynnis tumifrons	Skin	ASC 0.90	Thuy et al. (2014)
Etmopterus spp.	Skin	ASC 8.60	Sotelo et al. (2016)
Euthynms affinis	White muscle	ASC 1.06	Mathew and Hassan (1996)
Gadus morhua	Skin	ASC 20	Sadowska et al. (2003)
Galeus spp.	Skin	ASC 14.17	Sotelo et al. (2016)
Hemibagrus macropterus	Skin	ASC 16.8; PSC 28.0	Silva et al. (2014)

(Table Continued)

Species	Source	Yield (%) (ww)	References
Himantura sp.	White muscle	ASC 2.30	Mathew and Hassan (1996)
Katsuwonus pelamis	Spine	ASC 2.47; PSC 5.62	Yu et al. (2014)
	Skull	ASC 3.57; PSC 6.71	Yu et al. (2014)
	Bone	42.3	Nagai and Suzuki (2000)
Lagocephalus gloveri	Skin	PSC 54.3	Senaratne et al. (2006)
Larimichthys crocea	Scales	ASC 0.37; PSC 1.09 (d.w.)	Wu et al. (2015), Wang et al. (2013)
Larimichthys polyactis	Scales	ASC 0.42; PSC 1.14 (d.w.)	Wu et al. (2015)
Lateolabrax japonicus	Scales	PSC 41.0	Nagai et al. (2004)
	Skin	PSC 51.4	Kumar and Nazeer (2013)
	Bone	40.7	Nagai and Suzuki (2000)
	Fin	ASC 5.2	Nagai and Suzuki (2000)
Leiodon cutcutia	Skin	PSC 44.7	Kumar and Nazeer (2013)
Leucoraja naevus	Skin	ASC 7.85	Sotelo et al. (2016)
Lutjanus lutjanus	Bone	ASC 1.59; PSC 10.94	Silva et al. (2014)
	Skin	ASC 5.31; PSC 18.7	Whitehurst and Oort (2009)
Lutjanus vita	Skin	ASC 9.0; PSC 4.7	Whitehurst and Oort (2009)
Magalaspis cordyla	Bone	ASC 30.5; PSC 45.1	Kumar et al. (2011)
	Skin	ASC 17.3; PSC 22.5	Kumar and Nazeer (2013)
Merluccius hubbsi	Skin	ASC 6	Ciarlo et al. (1997)
Mugil cephalus	Scales	ASC 0.43	Thuy et al. (2014)
Nempterus japonicus	Scales and fins	ASC 6.9–22.0	Normah and Nur-Hani Suryati (2015)
Nempterus sp.	Skin	ASC 22.45; PSC 74.48	Whitehurst and Oort (2009)
Nezumia aequalis	Skin	ASC 14.73	Sotelo et al. (2016)
Otolithes ruber	Bone	ASC 27.6; PSC 48.6	Kumar et al. (2011)
	Skin	ASC 21.9; PSC 25.7	Kumar and Nazeer (2013)
Pagrus major	Scales	PSC 37.5	Whitehurst and Oort (2009), Nagai et al. (2004)
Pampus argenteus	White muscle	ASC 0.19	Mathew and Hassan (1996)

(Table Continued)

Species	Source	Yield (%) (ww)	References
Parupeneus heptacanthus	Scales	ASC 0.46; PSC 1.20 (d.w.)	Matmaroh et al. (2011)
Plecoglossus altivelis	Bone	53.6	Nagai and Suzuki (2000)
Pogonias cromis	Skin	ASC 2.3; PSC 15.8	Ogawa et al. (2003)
Pomadasys kaakan	Skin	PSC 439 mg/g	Aukkanit and Garnjanagoonchorn (2010)
	Skin (lyophilized)	0.90/g	Noitup et al. (2005)
Priacanthus tayenus	Skin	ASC 10.94	Kittiphattanabawon et al. (2005)
	Bone	ASC 1.19	Kittiphattanabawon et al. (2005)
Rastrelliger kanagurta	White muscle	ASC 0.38	Mathew and Hassan (1996)
Sardinella longiceps	White muscle	ASC 0.36	Mathew and Hassan (1996)
	Scales	ASC 1.25; PSC 3.0	Muthumari et al. (2016)
Saurida spp.	Scales	ASC 1.5	Thuy et al. (2014)
Scomber japonicas	Skin	PSC 49.8	Kumar and Nazeer (2013)
	Skin	49.8 type 1	Nagai and Suzuki (2000)
Scomberomorous niphonius	Skin	ASC 13.68; PSC 3.49	Li et al. (2013)
	Bone	ASC 12.54; PSC 14.27	Li et al. (2013)
Scophthalmus maximus	Skin	ASC 241.6 (per 1000 amino acids)	Sun et al. (2019)
Scyliorhinus canicula	Skin	ASC 11.60	Sotelo et al. (2016)
Sardine	Scales	PSC 50.9	Whitehurst and Oort (2009)
Sebastes mentella	Skin	ASC 47.5; PSC 92.2	Nalinanon et al. (2007)
	Skin	ASC 47.5; PSC 92.2	Kumar and Nazeer (2013)
Sillago sihama	White muscle	ASC 1.00	Mathew and Hassan (1996)
Silurus meridionalis	Defatted skin	23.14%; 78.57% (d.w.)	Xu et al. (2017)
Solea solea	Skin	ASC 0.1–0.175	Blidi et al. (2017)

(Table Continued)

Species	Source	Yield (%) (ww)	References
Syngnathus schlegeli	Skin	ASC 5.5; PSC 33.2	Silva et al. (2014)
Synodus foetens	Scales	ASC 0.79	Silva et al. (2014)
Takifugu rupripes	Skin	ASC 10.7; PSC 44.7	Silva et al. (2014)
Thunnus alalunga	Skin (lyophilyzed)	0.67/g	Noitup et al. (2005)
Thunnus albacores	Skin	PSC 27.1	Whitehurst and Oort (2009)
Thunnus orientalis	Skin	4.1 type 1	Han (n.d.)
Trachurus trachurus	Scales	ASC 1.51	Silva et al. (2014)
	Bone	43.5	Nagai and Suzuki (2000)
Trachurus japonicas	Scales	ASC 0.43–1.5	Thuy et al. (2014)
Trichiurus savala	White muscle	ASC 0.12	Mathew and Hassan (1996)

ASC, Acid-soluble collagen; PSC, pepsin-soluble collagen; dw, dry weight.

Collagen content of freshwater bony fishes (values in %; wet wt basis).

Species	Source	Yield (%) (ww)	References
Carassius carassius	Skin	PSC 40	He et al. (2019)
Catla catla	Skin	ASC 5.8; PSC 7.2	Mahboob (2015)
	Scales	ASC 3.9; PSC 5.6	Mahboob (2015)
	Fins	ASC 6.7; PSC 8.8	Mahboob (2015)
Channa striatus	Scales	ASC 1.44; PSC 2.94	Pamungkas et al. (2019)
Cirrhinus mrigala	Skin	ASC 4.7; PSC 6.5	Mahboob (2015)
	Scales	ASC 3.2; PSC 5.1	Mahboob (2015)
	Fins	ASC 5.7; PSC 7.7	Mahboob (2015)
Clarias batrachus	Skin	29.7	Tylingo et al. (2016)
Ctenopharyngodon idella	Skin	PSC 46.6	Zhang et al. (2007)
	Swim bladders	ASC 8.21	Zhang et al. (2010)
	Skin	PSC 28	He et al. (2019)
	Scales	PSC 27.33	He et al. (2019)
Cyprinus carpio	White muscle	ASC 0.49	Mathew and Hassan (1996)
Farbus spp.	White muscle	ASC 0.66	Mathew and Hassan (1996)

(Table Continued)

Species	Source	Yield (%) (ww)	References
Hemibagrus macropterus	Skin	ASC 16.8; PSC 28.0	Zhang et al. (2009)
H. molitrix	Skin	ASC (hydroxyproline and proline) 192/1000 residues	Zhang et al. (2009)
Hypophthalmichthys nobilis	Fins	PSC 2.0	Liu et al. (2012)
	Scales	PSC 1.1	Liu et al. (2012)
	Skin	PSC 17.5	Liu et al. (2012)
	Bones	PSC 1.3	Liu et al. (2012)
	Swim bladders	PSC 14.6	Liu et al. (2012)
Ictalurus punctatus	Skin	ASC 43.0	Tylingo et al. (2016)
	Skin	ASC 25.8; PSC 38.4	Kumar and Nazeer (2013)
Labes rohita	White muscle	ASC 0.28	Mathew and Hassan (1996)
Lates niloticus	Skin	ASC 58.7	Muyonga et al. (2004)
Megalops cyprinoides	White muscle	ASC 0.38	Mathew and Hassan (1996)
Oreochromis mossambicus	White muscle	ASC 0.47	Mathew and Hassan (1996)
Oreochromis niloticus	Skin	ASC 18.07	He et al. (2019)
Osteochilus vittatus	Skin	PSC 6.18	Junianto and Rizal (2018)
Pangasianodon hypophthalmus	Skin	ASC 5.1; PSC 7.7	Singh et al. (2011)
Pangasius sp.	Skin	ASC 4.27; PSC 2.27	Hukmi and Sarbon (2018)
Silurus meridionalis	Skin	23.14	Xu et al. (2017)
Systomus orphoides	Scales	ASC 0.43; PSC 0.60 (dw)	Aichayawanich and Parametthanuwat (2018)

ASC: Acid-soluble collagen; PSC, pepsin-soluble collagen; dw, dry weight.

Gelatin content (%) of marine fishes.

Species	Skin	Entire fish
Epinephelus sexfasciatus	68.5	3.7
Lutjianus argentimaculatus	55.2	1.8
Rastrelliger kanagurta	67.8	2.0
Pristipomodes typus	43.6	1.7

Source: Irwandi et al. (2009).

Gelatin content (%) of the skin of freshwater fishes.

Species	Gelatin
Pangasius pangasius	22.0
Oreochromis niloticus	21.9
Hemibagrus nemurus	21.3
Channa striata	20.3

Source: Ratnasari et al. (2013).

3.3.1.1.2   Applications of Fish Collagen and Gelatin

Wound healing and clinical importance: Fish collagens have been found reported to heal wounds resulting burns, grafting, ulcerations, and soon by preventing moisture and heat loss from the wounded tissue, and providing a microbial infiltration barrier. These collagens are also helpful in drug-delivery systems (Silva et al., 2014).

  Anticancer activity of fish collagen: The type 1 collagen derived from the skin of *Thunnus orientalis* has shown anticancer activity against human liver cancer cell line (HepG2) with 22% reduction of cell growth (Han, n.d.).

3.3.1.1.3   Therapeutic and Industrial Applications of Fish Gelatin

Fish gelatin is utilized as edible film, food emulsifier, thickener, and stabilizer in which the gelatin improves the mechanical and functional properties of gels. Desserts are made from fish gelatins by enhancing the concentration of gelatin (Zhou and Regenstein, 2007). As fish gelatins have low gelling temperatures, they are used in the preparation of pharmaceutical and medical products. For example, fish gelatin may be of great use as an ingredient in drug tablets, in the microencapsulation of vitamins, and other therapeutical additives such as azoxanthine. Fish gelatin soft capsules find application in nutrition supplements. In foods formulated for obese patients, fish gelatin is also recommended for reducing the fat levels (Liu et al., 2015; Karim and Bhat, 2008, 2009).

3.3.1.1.4   Mosquito Larvicidal Uses of Collagen

The collagen derived from the scales of *Sardinella longiceps* showed larvicidal activity against the larvae of *Aedes aegypti*. The percentage of mortality was increased with increases in collagen concentration. The LC50 value of *A. aegypti* mosquito larvae was observed at 90.788 L/100 mL (Muthumari et al., 2016). The percentage values of larvicidal activity against *A. aegypti* are given below:

Larvicidal activity of PSC collagen extracted from *S. longiceps* fish scales against *A. aegypti*.

Conc./100 mL	% of Mortality
20	20
40	25
60	35
80	50
100	60

Source: Muthumari et al. (2016).

3.3.1.2   FISH PROTEIN HYDROLYSATES AND THEIR BIOACTIVE PEPTIDES

3.3.1.2.1   Fish Protein Hydrolysates (FPHs) from Fish Processing Wastes

The fish processing wastes such as head, skin, viscera, frame, bone, and so on can be used in the production of fish protein hydrolysates (FPHs) by enzymatic hydrolysis and addition of bacterial or digestive proteases.

In the production of FPH the following are the important sources:

1. Processing wastes (such as heads, viscera, frames, skin, and trimmings) of black scabbardfish, *Aphanopus carbo*
2. Visceral and head wastes of *Sardinella aurita*
3. Surimi wastes (frame, bone, skin, and refiner discharge) from threadfin bream (*Nemipterus* spp.)
4. Frame from yellowfin sole (*Limanda aspera*), Alaska pollock (*Theragra chalcogramma*) and hoki (*Johnius belengerii*), and backbones from tuna and Atlantic cod (*Gadus morhua*)
5. Liquid effluent such as cooking juice from tuna

3.3.1.2.2   Production of FPHs

Enzymatic or chemical reactions lead to the production of FPH. Among the different sources of enzymes, bacterial, and fungal proteases are commonly employed as they have efficient catalytic activities and are more stable at high pH range and temperatures. Further, fungal proteases have been reported to display a broad substrate specificity which is responsible for more pronounced hydrolysis of protein. The abstract of processes in the production of FPH is given below:

Fish wastes (head, viscera, skin, bone, or scales) + collagen and gelatin (derived from skin, bone, and scales) > Enzyme addition > Hydrolysis with parameters viz. pH, temp., and time) > Enzymatic inactivation > Filtering/Centrifugation > supernatant > Drying > FPH powder

Enzymatic hydrolysis: This process is carried out with proteolytic enzymes, namely, endopeptidases and exopeptidases. Production of FPH is largely influenced by many factors, such as the composition of raw material, type of enzyme used, hydrolysis conditions, and degree of hydrolysis (DH).

FPH production from fish wastes with fish proteases.

Protease	Raw material
Tuna pyloric caeca proteases	*Gadus morhua* frame protein
Pepsin, mackerel intestine proteases	*Limanda aspera* frame
Mackerel intestine proteases	*Theragra chalcogramma* frame
Sardine viscera protease	*Sardinella aurita* heads and viscera

Source: Benjakul et al. (2014).

Fish enzymes especially from viscera, are also more suitable for preparing FPH from fish wastes. The list of commercial proteases used in the production of FPH is given below:

Protease	By-product source
Protamex	*Salmo salar* frames
Alcalase	*Clupea harengus* head, gonad
Alcalase	*Salmo salar* head
Flavorzyme	*Gadus morhua* viscera, backbone
Alcalase	*Oncorhynchus nerka* head
Orientase	*Thunnus tonggol* cooking juice
Alcalase	*Thunnus albacares* viscera
Protamex	*Aphanopus carbo* heads, viscera, frames, skin, trimmings
Alcalase, papain	*Cirrhinus mrigala* fish egg
Alcalase, trypsin, protamex	*Gadus chalcogrammus* frame muscle
Alcalase	*Nemipterus* spp. bone, skin

Source: Benjakul et al. (2014).

3.3.1.2.3   Bioactivities of Fish Protein Hydrolysates (FPH) Derived from Fish Wastes

i) **Antioxidant peptides from FPH of fish wastes and their amino acid sequences.**

By-product source	Amino acid sequence
Alaska Pollack frame	Leu–Pro–His–Ser–Gly–Tyr
Hoki skin gelatin	His–Gly–Pro–Leu–Gly–Pro–Leu
Tuna backbone	Val–Lys–Ala–Gly–Phe–Ala–Trp–Thr–Ala–Asn–Gln–Gln–Leu–Ser
Hoki frame	Glu–Ser–Thr–Val–Pro–Glu–Arg–Thr–His–Pro–Ala–Cys–Pro–Asp–Phe–Asn
Tuna cooking juice	Pro–His–His–Ala–Asp–Ser
Yellow fin sole frame	Arg–Pro–Asp–Phe–Asp–Leu–Glu–Pro–Pro–Tyr
Tuna cooking juice	Pro–Val–Ser–His–Asp–His–Ala–Pro–Glu–Tyr
	Pro–Ser–Asp–His–Asp–His–Glu
	Val–His–Asp–Tyr
Tuna dark muscle by-product	Pro–Met–Asp–Tyr–Met–Val–Thr
	Leu–Pro–Thr–Ser–Glu–Ala–Ala–Lys–Tyr
Horse mackerel skin	Asn–His–Arg–Tyr–Asp–Arg
Sardine (head, viscera)	Leu–His–Tyr

Source: Benjakul et al. (2014).

ii) **Antioxidative activity of FPH derived from tuna cooking juice**
Tuna cooking juice is a protein-rich by-product containing nearly 4% water-soluble protein and is discarded as drainage in commercial canned tuna factories. Further, the hydrolysate obtained from this juice using Protease XXIII from *Aspergillus oryzae* yielded seven anti-oxidative peptides which yielded 1,1-diphenyl-2-picryl-hydrazyl (DPPH)-radical scavenging ability at 82.2% (Jao and Ko, 2002). Furthermore, treated and processed juice can also be used as a functional additive to food stuffs (Gamarro et al., 2013). Tuna cooking juice is processed in two steps, namely, filtration of juice to remove the debris and subsequent precipitation with trichloroacetic acid. The precipitated protein is freeze-dried and resuspended in distilled water to constitute a 4% tuna protein solution. The rehydrated protein solution is irradiated by a cobalt-60 irradiator.

iii) **ACE inhibiting peptides from FPH of fish wastes and their amino acid sequences**

By-product sources	Amino acid sequence
Alaska pollock skin	Gly–Pro–Leu, Gly–Pro–Me
Yellowfin sole frame protein	Met–Ile–Phe–Pro–Gly–Ala–Gly–Gly–Pro–Glu–Leu
Salmon protein	Val–Leu–Trp, Val–Phe–Tyr, Leu–Ala–Phe
Tuna frame protein	Gly–Asp–Leu–Gly–Lys–Thr–Thr–Thr–Val–Ser–Asn–Trp–Ser–Pro–Pro–Lys–Try–Lys–Asp–Thr–Pro
Sardinelle heads and viscera	Leu–His–Tyr, Leu–Ala–Arg–Leu, Gly–Gly–Glu, Gly–Ala–His, Gly–Ala–Trp–Ala, Pro–His–Tyr–Leu, Gly–Ala–Leu–Ala–Ala–His

Source: Benjakul et al. (2014); Ewart et al. (2009).

iv) **Antimicrobial activity of FPHs:** Almost all fish antimicrobial peptides derived from the protein hydrolysates of fish wastes have antibacterial or bacteriostatic properties against both Gram-negative and Gram-positive strains (Najafian and Babji, 2012). Peptide fractions from alcalase hydrolysates of tuna gelatins had antimicrobial and antioxidant and properties (Gómez-Guillén et al., 2010). The wastes of mudfish, *Misgurnus anguillicaudatus* has yielded a novel antimicrobial peptide, misgurin with the amino acid sequence of Arg–Gln–Arg–Val–Glu–Glu–Leu–Ser–Lys–Phe–Ser–Lys–Lys–Gly–Ala–Ala–Ala–Arg–Arg–Arg–Lys. This peptide has been reported to display significant antimicrobial activity in vitro (Park et al., 1997).

v) **Calcium-binding activity (calcium absorption) of FPHs:**
Oligopeptides derived from the bone of hoki (*Macruronus novaezelandiae*) using tuna intestine crude enzyme exhibited high affinity to calcium. It is also reported that the bone peptide II of hoki inhibited the formation of insoluble Ca salts (Jung et al., 2006a). Fish bone oligophosphopeptide therefore has the potential to be used as a nutraceutical to increase the absorption of Ca (Simpson, 2012). Bones of Atlantic salmon and Atlantic cod have been reported as potential Ca sources in functional foods or as supplements. Young men fed with meals containing cod and salmon bones had recorded Ca absorption at 21.9% and 22.5%, respectively (Malde et al., 2010).

vi) **Anticoagulant activity of FPH:** Peptides derived from the fish frame muscle have displayed in-vitro antiplatelet and anticoagulant properties (Rajapakse et al., 2005). The protein hydrolysates prepared from goby muscle by treatment with various bacterial alkaline proteases showed anticoagulant activity with significant prolongation of both the thrombin time and the activated partial thromboplastin time. The hydrolysate generated by the crude protease from *Bacillus licheniformis* NH1 displayed the highest anticoagulant activity (Nasri et al., 2012a).

vii) **Antitumor activity:** Proteins hydrolysates from marine fish are a potential source of active biopeptides that can inhibit the growth of tumor cells. The small protein syngnathusin isolated form the whole body of *Syngnathus acus* was found to significantly inhibit the growth of A549 and CCRF-CEM cells (Khora, 2013). It could inhibit the growth of S180 tumor transplanted in mice; therefore, syngnathusin can potentially serve as an antineoplastic agent (Kristinsson, 2014). Hepcidin TH15, an antimicrobial peptide synthesized from tilapia, showed antitumor activity against several tumor cell lines. TH15 inhibited the proliferation of tumor cells (Chang et al., 2011).

3.3.1.2.4 Industrial Applications of FPH

3.3.1.2.4.1 Pharmaceutical and Nutraceutical Applications

Even though FPHs exhibit many health benefits, very few commercial products containing FPHs are available as human food.

Pharmaceutical (disease-preventing) activities of fish protein–derived peptides from fish wastes.

| Antioxidant activity |
| Anticancer activity |
| Antitumor activity |
| Anithypertensive activity |
| Antimicrobial activity |

FPHs have also been reported as a suitable source of protein for human nutrition because of their balanced amino acid composition and positive effect on gastrointestinal absorption (Benjakul et al., 2014).

Nutraceutical (health promoting) activities of fish protein derived peptides from fish wastes

| Anti-inflammatory activity |
| Anticoagulant activity |
| Immunomodulatory activity |
| Lipid-lowering activity |
| Calcium-binding activity |

Source: Benjakul et al. (2014).

Commercial nutraceuticals from FPHs.

Product	Source	Applications
Seacure®	Proteins of white fish	Dietary supplement
Amizate®	Proteins of Atlantic salmon	Sports nutrition
Stabilium®200	*Molva dypterygia*	Nutritional support
Protizen®	Proteins of white fish	Stress and symptoms
Vasotensin®	*Sarda orientalis*	Healthy vascular function
Peptace®	*Sarda orientalis*	ACE inhibition
Nutripeptin®	Cod fish	Weight management
Liquamen®	*Molva molva*	Dietary supplement
Molval®	*Molva molva*	Dietary supplement
Seagest®	Proteins of white fish	Intestinal health

Source: Ghaly et al. (2013).

3.3.1.3   FISH SAUCE AND ITS BIOACTIVITIES

Salt fermentation of processing wastes of salmon, sardine, and anchovy has yielded fish sauce of high quality. In the production of fish sauce, the said wastes are first treated with salt at 3:1 and are kept for about six months at a temperature of 30°C. By this process, an amber protein solution results and is drained from the bottom of the tank. Fish sauce with its rich amino acids is utilized as a condiment in vegetable dishes. Fermented fish sauce has also shown ACE inhibitory activity and insulin stimulating activity (Ichimura et al., 2003).

3.3.2   FISH OILS

Processing by-products of bony fishes such as tuna, salmon, sardine, white fish, dogfish, catfish, pollock, Atlantic herring, mackerel, horse mackerel, and so on are the potential sources for fish oil production. These fish oils are rich in n-3 PUFA family of α-linolenic acid, eicosapentaenoic acid, docosapentaenoic acid, and docosahexaenoic acid. Fish oils are recommended for patients with various diseases such as cancer, blood pressure, and inflammation.

Fatty acid composition of fish oil.

Fatty acid	%
14:0	3.4
15:0	0.7
16:0	16.1
16:1 n-7	0.8
16:1 n-9	5.3
17:0	1.1
17:1	0.5
18:0	4.6
18:1 n-7	2.3
18:1 n-9	12.7
18:2 n-6	1.4
18:3 n-3	0.6
18:3 n-6	1.8

(Table Continued)

Fatty acid	%
20:1 n-9	1.2
20:4 n-6	2.1
20:5 n-3	11.1
22:5 n-3	1.9
22:5 n-6	1.9
22:6 n-3	29.1

Source: Hosomi et al. (2013).

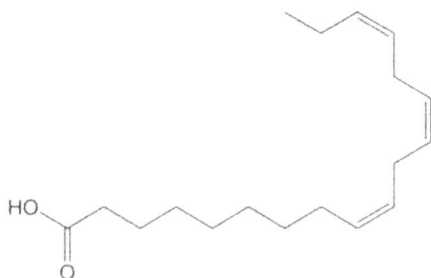

α-Linolenic acid

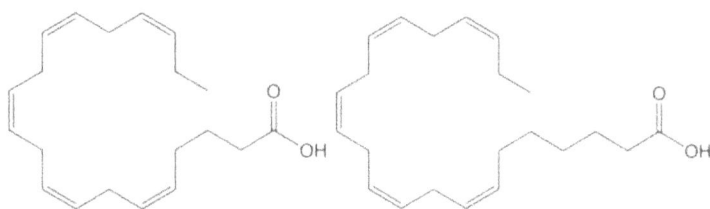

Eicosapentaenoic acid

Dcosapentaenoic acid

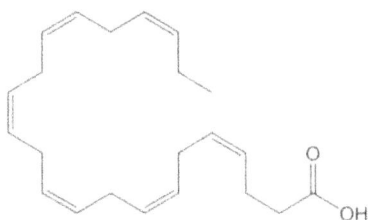

Docosahexanenoic acid

Fatty acid composition (%) of *Brevoortia tyrannus, Salmo gairdneri, Sardine pilchardus, Engraulis encrasicholus, and Katsuwonus pelamis.*

Component	B. tyrannus	S. gairdneri	S. pilchardus	E. encrasicholus	K. pelamis
SFA	30.5	18.4	27.2	40.4	38.4
MUFA	24.9	54.5	19.7	28.4	7.3
PUFA	27.6	24.8	28.4	29.4	45.7

SFA, Saturated fatty acids; MUFA, mono-unsaturated fatty acids; PUA, polyunsaturated fatty acids.

Source: Shahidi (2006).

Fatty acid composition (%) of oils from *Oncorhyncus gaorbuscha, Ictalurus punctatus, Gadus morhua, Clupea harengus,* and *Mallotus villosus.*

Component	O. gaorbuscha	I. punctatus	G. morhua	C. harengus	M. villosus
SFA	22.49	33.68	28.75	28.19	15.2
MUFA	39.18	32.68	20.97	36.39	61.9
PUFA	32.53	26.00	46.93	35.42	15.3

Source: Shahidi (2006).

Fatty acid composition (%) of oils from certain organs of *Oncorhyncus gaorbuscha, Ictalurus punctatus,* and *Gadus morhua.*

Component	O. gaorbuscha		I. punctatus	G. morhua
	Head	Viscera	Viscera	Liver
SFA	19.38	20.69	32.90	19.17
MUFA	41.73	33.83	38.44	49.07
PUFA	31.38	40.35	28.62	29.29

SFA, Saturated fatty acids; MUFA, mono-unsaturated fatty acids; PUA, polyunsaturated fatty acids.

Source: Shahidi (2006).

Omega-3 PUFAs (EPA and DHA) (%) of freshwater fish and marine bony fish—a comparison.

Fish	EPA	DHA
Freshwater fish	5–13	1–5
Marine fish	3–18*	2–13*
1. Pacific anchovy	18	11
2. Capelin	9	3
3. Mackerel	8	8
4. Herring	3–5	2–3
5. Sardine	3	9–13

*Average of 1–5.

Source: Shahidi (2006).

Omega-3 fatty-acids (EPA and DHA) content (g/100 g) in marine bony fishes.

Fish species	EPA	DHA
Chlorophthalmus agassizi	3.3	14.6
Sparus aurata	0.02	0.1
Spicara smaris	9.3	18.4
Merluccius merluccius	4.6	14.5
Helicolenus dactylopterus	3.7	17.0
Gadus morhua	16.4	27
Solea solea	19	12.3
Platichthys flesus	9.5	6.7
Sardina pilchardus	10.7	20.8
Dicentrarchus labrax	0.3	0.4
Engraulis encrasicolus	10.2	20.1
Hilsa macrura	11.8	6
Sarda sarda	8.1	21.5
Alosa immaculata	5.5	18.1
Salmo salar	4.2	6.8
Spicara smaris	9.3	18.4

EPA, Eicosapentaenoic acid; DHA, docosahexaenoic acid.
Source: Olgunoglu (2017).

Omega-3 fatty-acids (EPA and DHA) content (g/100 g) in freshwater bony fishes.

Species	EPA	DHA
Alburnus mossulensis	1.5	4
Barbus rajonorum	13.4	26.9
Carasobarbus luteus	6.9	12
Leuciscus lepidus	6.5	27.1
Acanthobrama marmid	8	10.8
Cyprinion macrostomus	20.2	22.2
Liza abu	0.7	0.7
Cyprinus carpio	4.8	6.7
Barbus grypus	3.2	12.8
Silurus triostegus	2.9	4.2
Mastacembelus mastacembelus	1.8	9.4
Pangasius pangasius	2.5	0.2
Oncorhynchus mykiss	7.2	5.4
Pseudoplatystoma corruscans	0.5	2.8
Piaractus mesopotamicus	0.7	1.9
Chalcalburnus tarichi	5.3	8.1
Esox lucius	8.1	22.7

(Table Continued)

Species	EPA	DHA
Clarias gariepinus	4.7	8.9
Oreochromis niloticus	0.6	5.3
Pangasianodon hypophthalmus	0.8	2.5
Anguilla anguilla	3.4	2.3
Vimba vimba tenella	5.3	6.7

EPA, Eicosapentaenoic acid; DHA, docosahexaenoic acid (22:6n-3).
Source: Olgunoglu (2017).

Finfish groups based on their omega-3 fatty-acids (ω-3 fatty acids or n-3 fatty acids) content.

Low (<0.5 g/100 g)	Medium (0.5–1 g/100 g)	High (>1 g/100 g)
Cod, Haddock, Pollack	Halibut	Mackerel
Grouper	Pike	Sablefish
Most flatfishes	Red snapper	Salmon (most species)
Perch	Swordfish	Bluefin tuna
Snapper	Trout	Whitefish
Carp	Bass	Anchovy
Catfish	Bluefish	Herring
Tilapia	Whiting	–

Source: Caballero (2009).

3.3.2.1   HEALTH BENEFITS FISH OILS

The pharmaceutical and nutraceutical potentials of omega-3 fatty acids of fish oils are given below:

(i)	Prevent of coronary artery disease
(ii)	Protect against arrhythmias
(iii)	Reduce blood pressure
(iv)	Beneficial for diabetic patients
(v)	Fight against manic-depressive illness
(vi)	Reduce symptoms in asthma patients
(vii)	Protect against chronic obstructive pulmonary diseases
(viii)	Alleviate symptoms of cystic fibrosis
(ix)	Improve the survival of cancer patients
(x)	Prevent relapses in patients with Crohn's disease
(xi)	Prevent autoimmune disorders
(xii)	Influence on mental health and psychiatric illness

Source: Shahidi et al. (2019).

3.3.2.2   PHARMACEUTICAL AND NUTRACEUTICAL APPLICATIONS OF FISH OILS

For pregnant and lactating women: Greenberg et al. (2008) reported that fish oil–based n-3 fatty acids exhibited pronounced properties on several pregnancy outcomes in women, including such as duration of gestation and infant size at birth; and preeclampsia (pregnancy complication including signs of damage to other organs like liver and kidneys), depression, cognition, and immunologic function.

Pharmaceutical importance: Among fish oils, the tuna oil which is a rich source of DHA has a number of health-related applications, including additive in infant formulas. Tuna oil–fed infants exhibited significant improvement in visual acuity. Cod and halibut liver oils are rich in n-3 PUFA and vitamins A and D and vitamins A and D, respectively. Supplementation of a regular diet supplemented with cod liver oil has shown beneficial effects for patients of cardiovascular and thrombotic diseases (Huang et al., 2018).

Food applications of omega-3 oils: Omega-3 oils find applications in bread/hard bread, cereals, noodles, pasta, and cakes. Further, these fatty acids are also used in milk, fortified, juices, tuna burger, eggs, canned tuna steak, and so on (Shahidi, 2002).

3.3.3   FISH BONES

Fish bones are rich in minerals such as calcium and phosphorus and collagen proteins. Apart from the latter, some special carbohydrate and lipids are also found in these bones. Fish bones could be used as a raw material for several health products. Fish frame and fish bone are sold as leisure snacks in Chinese and Korean markets (Shahidi et al., 2019).

Chemical composition of fish bones: Fish bones of different species vary significantly in their protein and lipid contents. Most pronounced difference however lies in lipid liver in the bones of different species. Studies relating to the chemical composition of cod (*Gadus morhua*), saithe (*Pollachius virens*), blue whiting (*Micromesistius poutassou*), salmon (*Salmo salar*), trout (*Salmo trutta*), herring (*Clupea harengus*), mackerel (*Scomber scombrus*), and horse mackerel (*Trachurus trachurus*) showed that fatty fish, namely, salmon, trout, herring, and mackerel which store lipid in the muscle have a much higher content of lipids in bones (318 g/kg) compared to the lean fish (blue whiting, cod, and saithe), the livers of which contain lipid at 20.6 g/kg (Toppe et al., 2007). The proximate and mineral composition of bones from the aforesaid species is given below:

Protein and lipid contents of fish bones (g/kg).

	Cod	Saithe	Blue whiting	Salmon	Trout	Herring	Mackerel	Horse mackerel
Protein	357.8	335.9	418.0	292.0	314.0	301.2	261.3	270.2
Lipid	11.4	14.9	49.1	381.2	343.7	266.7	471.8	226.1

Source: Toppe et al. (2007).

Mineral content of fish bones from dry fish.

Mineral	Cod	Saithe	Blue whiting	Salmon	Trout	Herring	Mackerel	Horse mackerel
Calcium (g/kg)	190	199	170	135	147	197	143	233
Phosphorous (g/kg)	113	108	87	81	87	95	86	111
Magnesium (g/kg)	3.0	3.0	3.2	2.2	2.4	2.9	2.6	3.6
Sodium (g/kg)	7.7	7.1	4.6	5.7	5.8	7.8	6.5	7.1
Potassium (mg/kg)	5.2	4.9	2.6	8.2	7.7	7.7	6.7	4.4
Iodine (mg/kg)	3.7	2.6	1.4	2.7	2.5	3.6	2.2	2.1
Iron (mg/kg)	49	44	135	32	32	72	73	56
Copper (mg/kg)	1.0	1.2	3.0	0.9	0.9	0.8	2.2	0.5

Source: Toppe et al. (2007).

3.3.3.1 APPLICATIONS OF FISH BONE MINERALS

Fish bone minerals as functional food and nutraceutical: In experimental rats fed with high calcium-containing tuna bone powder, there was a considerable increase in their mineral density and improvement in maternal bone microstructure. This suggests that the fish bone powder may act as calcium-fortified food supplements (Suntornsaratoon et al., 2018).

Fish bone minerals in biomedical industry: Fish bone–derived hydroxyapatite may be used as an alternate for synthetic hydroxyapatite (Granito et al., 2018). In this regard, the bones of salmon and rainbow trout may serve as a source of hydroxyapatite toward formation of mineralized tissue and proliferation of osteoblast and (Shi et al., 2018).

3.3.4   FISH ENZYMES

Proteases: Among the fish wastes, the intestines are the potential source of enzymes. Of these enzymes, the proteases are often used in the decalcification and curing of seafood products, for safe removal of fish skin and for tenderizing the fillets (Anon., https://www.eurofishmagazine.com/sections/fisheries/item/445-fish-entrails-and-processing-waste-as-a-raw-material).

Lipases: These enzymes derived from the viscera of Alaska pollock are known to exert significant effects on the quality of postharvest seafoods (Kurtovic et al., 2009).

3.3.4.1   BIOACTIVITIES OF ENZYMES EXTRACTED FROM FISH WASTES

ACE inhibitory activity.

Fish waste	Enzyme	ACE inhibitory activity (%)
Sardine wastes	Protease-K	47.4
	Alcalase	43
	Visceral enzyme	63.2
	Chymotrypsin	13.2
Yellowfin sole frame	Chymotrypsin	34.5–68.8
Pinperch frame	Papain	69
Tilapia wastes	Cryotin	62–71
	Flavourzyme	66–73

Source: Gamarro et al. (2013).

3.3.4.2   APPLICATIONS OF DIGESTIVE ENZYMES EXTRACTED FROM FISH

Pepsin: Pepsins from fishes such as Atlantic cod, sea perch, and orange roughly are largely used in the production of caviar (salt-cured roe) (Guérard et al., 2005).

Protease: The fish proteases find applications in silage, fish sauce, and fermentations (Squires et al., 1986).

Transglutaminase: These fish enzymes are often used in food products such as meat products, and surimi and other seafoods (Shahidi, 2006).

Lipase: The lipase derived from the seabass liver is mainly utilized in the defattening of fish skin (Oscarsson and Hurt-Camejo, 2017).

3.3.5   FISH POWDER FROM BONE AND FRAME WASTES

Dried fish wastes and underutilized fish are the potential sources for the production of fish powder which contains micronutrients of nutraceutical importance, such as phosphorus and calcium (Abbey et al., 2017).

3.3.5.1   PROCESSES INVOLVED IN FISH POWDER PREPARATION

i)   By-products (trimmings, gills and frames > Thawing > Washing > Drying (55°C for 8 h) > Milling > Fish powder

ii)   Underutilized fish > Thawing > Descaling > Degutting > Washing > Drying (55°C for 8 h)> Milling > Fish powder

Calcium content of frame and bone powder of fishes.

Calcium content (g/100 g) of fish bone powder and fish frame powder		
Fish	Fishbone powder	Fish frame powder
Tuna	38.16	24.56
Cod	–	19.0
Trout	–	14.7
Salmon	22.3	13.5
Saithe	–	19.9
Blue whiting	–	17.0
Herring	–	16.1
Oil sardine	32.73	–
Mackerel	–	14.3
Ribbon fish	27.81	–
Catfish	21.00	–
Snapper	24.40	–
Pollack	38.27	–

Source: Nemati et al. (2017).

Proximate composition and mineral contents of tuna bone powder.

Proximate composition	
Protein (g/100 g)	33.4
Fat (g/100 g)	11.0
Minerals	
Sodium (g/100 g)	0.5
Potassium (mg/100 g)	0.5
Calcium (g/100 g)	24.6
Phosphorus (g/100 g)	14.6
Magnesium (g/100 g)	0.3
Copper (mg/100 g)	0.1
Iron (mg/100 g)	4.3
Zinc (mg/100 g)	0.04

Source: Nemati et al. (2017).

Amino acids of tuna bone powder (g/100 g).

Alanine	7.9
Arginine	7.9
Aspartic acid	5.1
Cystine/Cicteine	0.1
Glycine	18.2
Histidine	2.4
Hydroxyprolin	7.9
Isoleucine	1.8
Lysine	3.0
Mehtionine	2.5
Phenylalanine	2.6
Proline	8.9
Serine	4.8
Threonine	3.7
Tryptophan	0.4
Tyrosine	2.8
Valine	3.0

Source: Nemati et al. (2017).

Fatty acid profile (g/100 g) of tuna bone powder.

SFA	28.9
MUFA	47.4
PUFA	23.6
PUFA (n-3)	21.4
PUFA (n-6)	1.5
PUFA/SFA	0.8

SFA, Saturated fatty acids; MUFA, mono-unsaturated fatty acids; PUA, polyunsaturated fatty acids.
Source: Nemati et al. (2017).

Tuna flavor powder: The tuna precooking juice which is normally discarded from canning factories is a potential raw material in the production of the tuna flavor powder. This juice is centrifuged and concentrated. To this concentrate, maltodextrin (DE 9) which is a polysaccharide serving as food additive is added and then at 180°C. The tuna flavor powder thus obtained is a product in human nutrition (Gamarro et al., 2013).

3.3.6   DRIED COD HEADS

The dried cod head is known for its nutraceutical properties due to its rich protein, mineral, and PUFA especially omega-3 fatty acids and is fit human consumption. Its protein has been reported to range from 54.1–60.5%; and PUFA at 37.8% (Salaudeen, 2014). While this by-product is largely produced in Norway and Iceland, its important market is Nigeria where it is sold in the trade name of "okporoko" which is utilized in soups and stews (Anon., https://www.eurofishmagazine.com/sections/fisheries/item/445-fish-entrails-and-processing-waste-as-a-raw-material).

3.3.7   ROE AND CAVIAR

Processed fish roe, which is a delicacy in many countries, is obtained by salting and drying the fresh fish roe. Fish roe is eaten either as roe sac or as individual eggs known as caviar. In the Mediterranean region, roe prepared from grey mullet, tuna, and swordfish is a very popular product which serves as a starter menu. In Japan, roe largely obtained from fishes like salmon, trout, and Alaska pollock. Several species of fishes such as sturgeon, Pacific, trout, lumpfish, capelin, pike, herring, and flying fish yield quality caviar (Bledsoe et al., 2003).

3.3.8 FISH'S MILT AND LIVER

Milt (soft roe) which is the fried and salted male sexual organs with semen is also a delicacy and is utilized as a topping for pasta dishes. Milt produced from herrings, cod, tuna, and carp is popular in Russia, Japan, Sicily, and Romania, respectively (Anon., https://www.eurofishmagazine. com/sections/fisheries/item/445-fish-entrails-and-processing-waste-as-a-raw-material). The livers of fishes such as cod, haddock, and blue ling are also nutritionally important. In Norwegian Lofoten Islands, cod liver is consumed as a sauce. The fried fatty livers of anglerfish (*Lophius piscatoius*) is also a delicacy in many countries (Anon., https://www.eurofishmagazine.com/sections/fisheries/item/445-fish-entrails-and-processing-waste-as-a-raw-material).

3.4 CARTILAGINOUS FISHES

3.4.1 LIVER OIL

Methods of extraction of silky shark (*Charcarinus falciformis*) liver oil (Jayasinghe et al., http://www.fao.org/3/a-bm158e.pdf; Bligh and Dyer, 1959) **Wet rendering method:** In this method, the minced liver is mixed with water (20%) and is boiled for 30 min. The concentrate is then centrifuged to separate the oil.

   Alkali digestion method: In this method, 10% water is added to liver mince and is digested at 40–50% until the liver in liquefied. Then 2% sodium hydroxide is added to this mixture and heating is continued at 80°C with pH adjusted to 9. Excess alkali is washed with warm water and the oil is separated by centrifugation.

   Steam rendering method: In this method, minced liver is placed in a round bottom flask (1000 mL) and the steam is passed through the sample for 2 h. The resulting mixture is then centrifuged at 2000 rpm for 10 min to obtain oil.

Incubation method: By this method, the liver mince is kept in an incubator at 30°C for 48 h and the oil is separated by centrifugation.

Bligh and Dyer method: This method is very simple and it can be carried out in approximately 10 min; it is efficient, reproducible, and free from deleterious manipulations. In this method, the wet shark tissue is homogenized with a mixture of chloroform and methanol in such proportions that a miscible system is formed with the water in the tissue. Dilution with chloroform and water separates the homogenate into two layers, namely, the chloroform layer which is containing all the lipids and the methanolic layer containing all the nonlipids. A purified lipid extract is then obtained by isolating the chloroform layer.

Silage method: In this method, 3.5% (w/w) formic acid is added to 100 g minced liver and is mixed thoroughly for 10 min. The mixtures are kept for 24 h at room temperature (28.1°C) and the oil is then separated by centrifugation at 2000 rpm for 10 min.

Yield (%) of crude silky shark liver oil extracted with different techniques.

Wet rendering	Alkali digestion	Steam rendering	Incubation	Bligh and Dyer	Silage
33.1	25.5	41.7	43.7	63.3	44.3

Source: Jayasinghe et al. (http://www.fao.org/3/a-bm158e.pdf).

Species of sharks yielding liver oil (Kuang, http://www.fao.org/3/x3690e/x3690e1d.htm).

Alopias pelagicus	*Carcharhinus obscurus*
Alopias superciliosus	*Carcharhinus plumbeus*
Alopias vulpinus	*Carcharodon carcharias*
Carcharhinidae taurus	*Centrophorus acus*
Carcharhinus albimarginatus	*Centrophorus lusitanicus*
Carcharhinus altimus	*Centrophorus squamosus*
Carcharhinus amblyrhynchos	*Centroscymnus crepidater*
Carcharhinus brevipinna	*Centroscymnus owstonii*
Carcharhinus falciformis	*Centroscymnus plunketi*
Carcharhinus leucas	*Cetorhinus maximus*
Carcharhinus limbatus	*Cirrhigaleus barbifer*
Carcharhinus longimanus	*Dalatias licha*
Carcharhinus melanopterus	*Dasyatis pastinaca*

(Table Continued)

Deania calcea	*Negaprion acutidens*
Echinorhinus brucus	*Odontaspis ferox*
Galeocerdo cuvier	*Prionace glauca*
Galeorhinus galeus	*Pristiophorus nudipinnis*
Galeus glucas	*Pristis pectinata*
Hemipristis elongata	*Rhincodon typus*
Hexanchus griseus	*Rhynchobatus djiddensis*
Isurus oxyrinchus	*Sphyrna diplana*
Isurus paucus	*Sphyrna lewini*
Lamna ditropis	*Sphyrna malleus*
Lamna nasus	*Sphyrma tudes*
Mustelus antarcticus	*Sphyrna zygaena*
Mustelus canis	*Squalus acanthias*
Mustelus manazo	*Squalus cubensis*
Mustelus mustelus	*Squalus mitsukurii*
Mustelus schmitti	*Squatina aculeata*
Nebrius ferrugineus	*Triaenodon obesus*

Fatty acids of liver oils of *Mustelus mustelus*, *Squalus acanthias*, and *Rhinobatos cemiculus*.

Fatty acid content (%) of liver oils certain species of sharks			
Fatty acids (%)	***M. mustelus***	***S. acanthias***	***R. cemiculus***
SFA	41.4	41.8	45.8
MUFA	23.6	35.3	27.1
PUFA	34.9	22.6	27.0
EPA	6.6	7.6	6.8
DHA	20.4	11.9	12.2

Source: Achouri et al. (2017).

3.4.1.1 PHARMACEUTICAL AND NUTRACEUTICAL PRODUCTS OF LIVER OIL

The eicosapentaenoic acid of shark liver oil has been reported to be very effective for treatment of some cardiovascular diseases and it has a protective effect against thrombosis, atherosclerosis, and inflammatory diseases. The docosahexaenoic acid of this oil is also known to help in the brain development and in preventing skin disorders. The PUFA, arachidonic acid (C20:4 x-6) was the major one present at levels ranging from 2.02 (for *S. acanthias*) to 5.36% (for *R. cemiculus*). This fatty acid which is a precursor for prostaglandins helps in immune response and thromboxane which contributes to form blood clot by attachment to the endothelial tissue during wound healing. Arachidonic acid also plays an important role in the development of the brain, retina, and infantile growth.

3.4.1.2   SQUALENE AND ITS PHARMACEUTICAL APPLICATIONS

The lipids of livers of shark species typically consist of a mixture of hydrocarbons (mainly squalene and some pristane), diacyl and monoglyceryl ethers and triglycerides. Squalene is an important component present in high concentrations (up to 89%) in the liver oils of deep-sea sharks (Peyronel et al., 1984). Squalene is considered very important therapeutically owing to its health benefits.

It is known to prevent diabetes, hepatitis, heart disease, allergies, and so on, in humans. Rao and Acaya (1968) reported on the antioxidant activity and skin-care benefits of squalene which acts as an effective moisturizer, wrinkle remover, and wound healer. Squalene supplements have also been reported to act as potential anticancer, antiradiatory, and anti-oxidative agents. Squalene is also known to serve as an emulsifier in adjuvant formulations of virus vaccines, and as an immunologic adjuvant (Kim and Karadeniz, 2012).

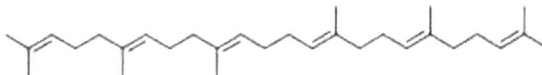

Squalene

3.4.2   CARTILAGE

Proximate composition (%) of shark cartilage.

Protein	14.0–19.2
Fat	0.2–1.4
Moisture	66.8–78.3
Ash	1.1–12.1

Source: Jeevithan et al. (2014).

Amino acid compositions of *Carcharhinus albimarginatus* cartilage.

Amino acid	Residues/1000 residues
Phe	19.51
Lys	22.34
His	9.98
Arg	50.96
Hyp	49.96
Asp	51.84
Thr	27.26
Cys	5.88
Val	25.63
Met	13.69
Ile	18.55
Leu	40.16
Tyr	15.13
Ser	35.62
Glu	86.45
Pro	93.61
Gly	310.47
Ala	122.87
Imino acid	143.58

Source: Jeevithan et al. (2014).

Collagen (type II) content (%) of cartilage of shark species.

Species	ASC	PSC
Chiloscyllium punctatum	1.3	9.6
Carcharhinus limbatus	10.4	10.3

ASC, Acid-soluble collagen; PSC, pepsin-soluble collagen.
Source: Silva et al. (2014).

Amino acid content (residues/1000 residues) of type II ASC and PSC collagens; and type II gelatin from *Carcharhinus albimarginatus* cartilage.

Amino acids	ASC	PSC	Gelatin
Asp	42.91	45.76	39.84
Thr	23.54	25.57	21.85
Ser	36.37	38.27	32.62

(Table Continued)

Amino acids	ASC	PSC	Gelatin
Glu	71.57	76.20	71.70
Pro	103.30	106.78	102.31
Gly	326.90	319.69	353.12
Hyp	47.50	49.25	51.28
Ala	133.92	132.63	140.45
Cys	3.93	4.26	3.74
Val	25.14	25.43	22.65
Met	10.68	13.54	12.34
Ile	21.19	21.81	16.94
Lys	30.93	29.37	27.35
His	9.33	9.38	8.15
Arg	55.05	49.87	53.14
Leu	30.01	29.95	25.16
Tyr	8.75	7.19	2.73
Phe	18.95	14.97	14.66
Imino acid	150.81	156.0	153.59

ASC, Acid-soluble collagen; PSC, pepsin-soluble collagen.
Source: Jeevithan et al. (2014).

3.4.2.1   *CHONDROITIN (CHONDROITIN SULFATE) AND COLLAGEN PEPTIDES FROM SHARK CARTILAGE*

Sources of chondroitin: The head of the blue shark, *Prionace glauca*, frame of the thornback ray, *Raja clavata*, and the small-spotted cat shark (*Scyliorhinus canicula*) are the major sources of chondroitin.

   Preparation of chondroitin and collagen peptides: In the preparation of chondroitin and collagen peptides, shark cartilage is ground with the enzyme, neutrase for 5 h. During this process, an enzyme to substrate ratio of 1800 U/g and solvent to material ratio of 10 mL/g are maintained. Hydrolysis of chondromucoids present in shark cartilage may also yield chondroitin. The peptides so extracted had the major amino acids such as glycine, proline, and hydroxyproline (Xie et al., 2014).

3.4.2.2 ANTIOXIDANT ACTIVITY OF CARTILAGE-DERIVED TYPE II COLLAGENS (ASC AND PSC); AND GELATIN OF CARCHARHINUS ALBIMARGINATUS

The type II collagens, namely, acid-soluble collagen (ASC) and pepsin-soluble collagen (PSC) and type II gelatin isolated from the cartilage of *Carcharhinus albimarginatus* possessed DPPH-radical scavenging activity and reducing power and the recorded values are given below (Jeevithan et al., 2014):

Antioxidant activity of cartilage-derived type II collagens and gelatin of *Carcharhinus albimarginatus*

Sample	DPPH-radical scavenging (%)	Reducing power*
ASC	20.1	0.2
PSC	24.8	0.2
Gelatin	16.6	0.2

*Absorbance at 700 nm.

Source: Jeevithan et al. (2014).

3.4.2.3 PHARMACEUTICAL USES OF CHONDROITIN

Chondroitin is utilized in the preservation of corneas meant for transplantation, and preparations with chondroitin or its sodium salt are used as adjuncts in ocular surgery. Chondroitin has also been reported to possess anticancer activity by limiting the growth of cancerous tumors and inhibiting the development of blood vessels in these tumors. Chondroitin has been found very useful for the treatment of patients with osteoporosis, hyper-lipidemia, arthritis, hemorrhoids, eczema, acne, ulcers, etc. (Shmerling, https://www.health.harvard.edu/blog/chondroitin-and-melanoma-how-worried-should-you-be-2018050913792).

3.4.2.4 THERAPEUTIC EFFECTS OF SHARK CARTILAGE

Anti-angiogenic and anticancer properties: The compound Neovastat (AE-941) which is a mixture of water-soluble components obtained from shark cartilage and shark cartilage extract exhibited anti-angiogenic activity by inducing apoptosis. AE-941 has also been reported to inhibit the activity of matrix metalloproteinases thereby controlling the metastatic potential of tumor cells (Patra and Sandell, 2012).

Anti-inflammatory effects: Neovastat-treated mice showed anti-inflammatory activity by significantly reducing the levels of matrix metalloproteinase (MMP9) activity in their bronchoalveolar lavage fluid, and reducing the VEGF and hypoxiainducible factor-2 expression in their lung tissue (Patra and Sandell, 2012).

Reduction of Kaposi Sarcoma: A human herpes virus 8 affected man was fed with shark cartilage and it was found that this treatment was found effective in reducing Kaposi Sarcoma lesion size, color, and small vessels in the above patient (Anon., http://www.ucdenver.edu/academics/colleges/pharmacy/currentstudents/OnCampusPharmDStudents/ExperientialProgram/Documents/nutr_monographs/Monograph-shark.pdf).

Others: In experimental rats, both the shark cartilage and shark liver oil have been found to protect these rats treated with dimethylhydrazine, a cancer causing agent (Akbulut and Akgül, 2018). Barman (https://www.researchgate.net/publication/257139523_Fish_Derived_Nutraceuticals_and_food_preservatives) reported that the shark cartilage is used in the treatment of atherosclerosis, blood vessel thrombosis, and in the reduction of cancer-related tumors and inflammation.

3.4.2.5   NUTRACEUTICAL PROPERTIES OF SHARK CARTILAGE

1. (Processed shark fin cartilage serves as a component in soups prepared chicken broth and herbs (Anon., https://sharkstewards.org/shark-finning/shark-finning-fin-facts/).
2. It is a dietary supplement in several countries like USA (Anon., https://www.fda.gov/food/dietary-supplements).
3. Soup prepared from the cartilage of jaw, fin, and head of sharks is a health supplement in countries like China (Anon., http://www.fao.org/3/x3690e/x3690e1d.htm).
4. Cartilage tablets or powder is fortified with various health enhancers (Vannuccini, 1999).

3.4.3   FINS

Shark fin soup is used in high-class Chinese cuisine. It was a delicacy once prepared exclusively for the Chinese emperors and nobility. Shark fin needles cooked with tasty ingredients are also served in restaurants in several south-east Asian countries. The nutritional composition of 100 g of dried sharks' fin needles is given below:

Proximate composition and minerals of shark fins.

Protein	83.5 g
Fat	0.3 g
Water	14.0 g
Ash	2.2 g
Iron	15.2 mg
Calcium	146.0 mg
Phosphorus	194.0 mg

Source: Anon. (http://www.fao.org/3/x3690e/x3690e1g.htm).

3.4.3.1   HEALTH BENEFITS OF SHARK FINS

Chinese medical books reported on the various health benefits of shark fins especially for the kidneys, lungs, and bones, and as an appetite enhancer (Anon., https://sharkstewards.org/shark-finning/shark-finning-fin-facts/).

3.4.4   SURIMI FROM THE SHARK MUSCLE

The shark meat obtained from the abundantly available but underutilized blue shark, *Prionace glauca* is converted into surimi for use in the preparation of a fish cake in the name of "kamaboko" in Japan (Ishimura and Bailey, 2013).

3.4.5   COLLAGENS FROM THE SKIN OF ELASMOBRANCHS

The skin of the elasmobranchs, namely, sharks and rays is also a potential source of collagen.

Collagen (type II, ASC) from the skin of certain elasmobranch species (wet wt basis).

Species	%
Etmopterus spp.	8.6
Galeus spp.	14.2
Scyliorhinus canicula	11.6
Leucoraja naevus	7.9
Carcharhinus limbatus	20.0
Chiloscyllium punctatum	9.4

ASC, Acid-soluble collagen; collagen is estimated based on the hydroxyproline content of the skin; ratio of HPro in collagen is 12.5 g of HPro/100 g of collagen (Edwards and O'Brien, 1980).

Source: Sotelo et al. (2016); Kittiphattanabawon et al. (2010c).

Amino acids of collagen (ASC) of elasmobranchs.

Amino acid	E	G	Sc	Ln	Cl	Cp
Alanine	127	114	100	99	118	105
Arginine	48	49	50	49	54	51
Aspartic a/asparagine	37	40	43	42	42	42
Cysteine	27	12	1	1	1	1
Glutamic a/glutamine	64	66	68	73	77	77
Glycine	318	337	338	334	317	318
Histidine	11	9	10	10	8	7
Isoleucine	18	14	13	17	20	18
Leucine	27	25	22	23	25	24
Lysine	25	26	27	27	28	29
Hydroxylysine	6	4	6	6	5	6
Methionine	22	17	14	14	14	12
Phenylalanine	16	14	15	14	14	14
Hydroxyproline	68	81	87	72	88	93
Proline	83	85	90	87	109	111
Serine	51	52	59	66	29	41
Threonine	20	25	27	32	21	23
Tyrosine	4	3	3	4	3	3
Valine	29	26	27	27	26	25
Imino acids	151	166	177	159	197	204

Etmopterus spp. *(E), Galeus* spp. *(G), Scyliorhynus canicula (Sc), Leucoraja naevus (Ln), Carcharhinus limbatus (Cl), and Chiloscyllium punctatum (Cp)* (residues/1000 residues).

Source: Sotelo et al. (2016), Kittiphattanabawon et al. (2010c).

References

Abbey, L.; Glover-Amengor, M.; Atikpo, M. O.; Atter, A.; Toppe, J. Nutrient Content of Fish Powder from Low Value Fish and Fish Byproducts. *J. Food Sci. Nutr.* **2017**, *5*, 374–379.

Achouri, N.; Smichi, N.; Kharrat, N.; Rmili, F.; Gargouri, Y.; Miled, N.; Fendri, A. Characterization of Liver Oils from Three Species of Sharks Collected in Tunisian Coasts: In Vitro Digestibility by Pancreatic Lipase. *J. Food Biochem.* **2017**, *42*, 1–9.

Adel, M.; Safari, R.; Soltanian, S.; Zorriehzahra, M. J.; Esteban, M. A. Antimicrobial Activity and Enzymes on Skin Mucus from Male and Female Caspian Kutum (*Rutilus frisii kutum* Kamensky, 1901) Specimens. *Slov. Vet. Res.* **2018**, *55*, 235–243.

Agustin, T. I.; Wahyu, S.; Yatmasari, E. Study on the Bioactive Compounds of Shark (*Prionace glauca*) Cartilage and Its Inflammatory Activity. *Int. J. PharmTech. Res.* **2016**, *9*, 171–178.

Ahmad, M.; Benjakul, S.; Nalinanon, S. Compositional and Physicochemical Characteristics of Acid Solubilized Collagen Extracted from the Skin of Unicorn Leatherjacket (*Aluterus monoceros*). *Food Hydrocolloids* **2010**, *24*, 588–594.

Aichayawanich, S.; Parametthanuwat, T. Isolation and Characterization of Collagen from Red Cheek Barb Scale (*Puntius orphoides*). In *Proceedings of the 2018 8th International Conference on Bioscience, Biochemistry and Bioinformatics*, 2018; pp 27–31.

Aissaoui, N.; Abidi, F.; Marzouki, M. N. ACE Inhibitory and Antioxidant Activities of Red Scorpionfish (*Scorpaena notata*) Protein Hydrolysates. *Int. J. Pept. Res. Ther.* **2015**, *55*. DOI:10.1007/s10989-016-9536-6.

Aissaoui, N.; Abidi, F.; Hardouin, J.; Abdelkafi, Z.; Marrakchi, N.; Jouenne, T.; Marzouki, M. N. Two Novel Peptides with Angiotensin I Converting Enzyme Inhibitory and Antioxidative Activities from *Scorpaena notata* Muscle Protein Hydrolysate. *Biotechnol. Appl. Biochem.* **2017a**, *64*, 201–210.

Aissaoui, N.; Marzouki, M. N.; Abidi, F. Purification and Biochemical Characterization of a Novel Intestinal Protease from *Scorpaena notata. Int. J. Food Prop.* **2017b**, *20*, 2151–2165.

Ajeeshkumar, K. K.; Asha, K. K.; Vishnu, K. V.; Remyakumari, K. R.; Shyni, K.; Reshma, J.; Navaneethan, R.; Linu, B.; Mathew, S. Proteoglycans Isolated from Bramble Shark Cartilage (*Echinorhinus brucus*) Inhibits Proliferation of MCF-7 Human Breast Cancer Cells by Inducing Apoptosis. *Biochem. Pharmacol.* **2017**, *6*, 7.

Akbulut, M. D.; Akgül, E. Determination of the Protective Effect of the Shark Cartilage and the Shark Liver Oil (Slo) against Formaldehyde and DMH Application Causing to Cancer and DNA Damage on Genetic Bases. *Int. J. Sci. Eng. Res. (IJSER)* **2018**, *9*, 51–55.

Akunne, T. C.; Okafor, S. N; Okechukwu, D. C.; Nwankwor, S. S.; Emene, J. O.; Okoro, B. N. Catfish (*Clarias gariepinus*) Slime Coat Possesses Antimicrobial and Wound Healing Activities, UK. *J. Pharm. Biosci.* **2016**, *4*, 81–87.

Alabssawy, A. N. Antimicrobial Activity of Tetrodotoxin Extracted from Liver, Skin and Muscles of Puffer Fish. *Lagocephalus sceleratus* Inhabiting Mediterranean Sea, Egypt. *Int. J. Cancer. Biomed. Res.* **2017,** *1,* 2–10.

Alolod, G. A. L.; Nuñal, S. N.; Nillos, M. G. G.; Peralta, J. P. Bioactivity and Functionality of Gelatin Hydrolysates from the Skin of Oneknife Unicornfish (*Naso thynnoides*). *J. Aquat. Food Prod. Technol.* **2019,** *28,* 1013–1026.

Al-Rasheed, A.; Handool, K. O.; Garba, B.; Noordin, M. M.; Bejo, S. K.; Kamal, F. M.; Daud, H. H. M. Crude Extracts of Epidermal Mucus and Epidermis of Climbing Perch *Anabas testudineus* and Its Antibacterial and Hemolytic Activities. *Egypt. J. Aquat. Res.* **2018,** *44,* 125–129.

Arakawa, T.; Tsumoto, K.; Kita, Y.; Chang, B.; Ejima, D. Biotechnology Applications of Amino Acids in Protein Purification and Formulations. *Amino Acids* **2007,** *33,* 587–605.

Athmani, N.; Dehiba, F.; Allaoui, A.; Barkia, A.; Bougatef, A.; Lamri-Senhadji, M. Y.; Nasri, M.; Boualga, A. *Sardina pilchardus* and *Sardinella aurita* Protein Hydrolysates Reduce Cholesterolemia and Oxidative Stress in Rat Fed High Cholesterol Diet. *J. Exp. Integr. Med.* **2015,** *5,* 47–54.

Atma, Y. Amino Acid and Proximate Composition of Fish Bone Gelatin from Different Warm-Water Species: A Comparative Study. *IOP Conf. Ser. Earth Environ. Sci.* **2017,** *58,* 012008.

Aukkanit, N.; Garnjanagoonchorn, W. Temperature Effects on Type I Pepsin-Solubilised Collagen Extraction from Silver-Line Grunt Skin and Its In Vitro Fibril Self-Assembly. *J. Sci. Food Agric.* **2010,** *90,* 2627.

Bagchi, D.; Preuss, H. G.; Swaroop, A. *Nutraceuticals and Functional Foods in Human Health and Disease Prevention*; CRC Press: Boca Raton, 2016.

Balasubramanian, S.; Revathi, A.; Gunasekaran, G. Studies on Anticancer, Haemolytic Activity and Chemical Composition of Crude Epidermal Mucus of Fish *Mugil cephalus*. *Int. J. Fish. Aquat. Stud.* **2016,** *4,* 438–443.

Bashir, K. M. I.; Young-Joo Park, Y.; An, J. H.; Choi, S.; Kim, J.; Baek, M.; Kim, A.; Sohn, J. H.; Choi, J. J. Antioxidant Properties of *Scomber japonicus* Hydrolysates Prepared by Enzymatic Hydrolysis. *Aquat. Food Prod. Technol.* **2018,** *27,* 107–121.

Batista, I.; Ramos, C.; Coutinho, J.; Bandarra, N. M.; Nunes, M. L. Characterization of Protein Hydrolysate and Lipids Obtained from Black Scabbardfish (*Aphanopus carbo*) by Products and Antioxidative Activity of the Hydrolysates Produced. *Process Biochem.* **2010,** 45, 18–24.

Benjakul, S.; Yarnpakdee, S.; Senphan, T.; Halldorsdottir, S. M.; Kristinsson, H. G. Fish Protein Hydrolysates: Production, Bioactivities, and Applications. In *Antioxidants and Functional Components in Aquatic Foods*; Kristinsson, H. G., Sivakumar Raghavan, S. Eds.; John Wiley & Sons: Hoboken, NJ, **2014**; pp 237–281.

Bernhardt, A.; Paul, B.; Gelinsky, M. Biphasic Scaffolds from Marine Collagens for Regeneration of Osteochondral Defects. *Mar. Drugs* **2018,** *16,* 91.

Bhargava, P.; Marshall, J. L.; Dahut, W.; Rizvi, N.; Trocky, N.; Williams, J. I.; Hait, H.; Song, S.; Holroyd, K. J.; Hawkins, M. J. A Phase I Pharmacokinetic Study of Squalamine, a Novel Antiangiogenic Agent, in Patients with Advanced Cancers. *Clin. Cancer Res.* **2001,** *7,* 3912–3919.

Binsi, P. K.; Viji, P.; Panda, S. K.; Mathew, S.; Zynudheen, A. A.; Ravishankar, C. N. Characterisation of Hydrolysates Prepared from Engraved Catfish (*Nemapteryx caelata*) Roe by Serial Hydrolysis. *J. Food Sci. Technol.* **2016,** *53,* 158–170.

Birkemo, G. A.; Lüders, T.; Andersen, Ø; Nes, I. F.; Nissen-Meyer, J. Hipposin, a Histone-Derived Antimicrobial Peptide in Atlantic Halibut (*Hippoglossus hippoglossus* L.). *Biochim. Biophys. Acta* **2003**, *1646*, 207–215.

Bkhairia, I.; Kolsi, R. B. A.; Ghorbel, S.; Azzabou, S.; Ktari, N.; Nasri, M. Anti-inflammatory, Antioxidant Activities and Fatty Acid Profile of Three Hydrolysates from *Liza aurata* By-product Influenced by Hydrolysis Degree. *Adv. Tech. Biol. Med.* **2019**, *7*, 268.

Bledsoe, G. E.; Bledsoe, C. D.; Rasco, B. Caviars and Fish Roe Products. *Crit. Rev. Food Sci. Nutr.* **2003**, *43*, 317–356.

Blidi, O.; Elomari, N.; Kamar-Zaman, Y.; Chakir, I.; Kaddafi, A.; Lebjar, N.; Ibrahimi, A.; Chokairi, O.; Barkiyou, M. Simplified Extraction and Characterization of Acetic Acid Solubilized Type I Collagen Derived from *Solea solea* Skin and Wistar Rat Tails for Biomedical and Biotechnological Applications. *J. Chem. Pharm. Res.* **2017**, *9*, 154–164.

Bligh, E. G.; Dyer, W. J. A Rapid Method of Total Lipid Extraction and Purification. *Can. J. Biochem. Physiol.* **1959**, 37, 911–917.

Borawska, J.; Darewicz, M.; Vegarud, G. E.; Iwaniak, A.; Minkiewicz, P. Ex Vivo Digestion of Carp Muscle Tissue—ACE Inhibitory and Antioxidant Activities of the Obtained Hydrolysates. *Food Funct.* **2015**, *6*, 211–218.

Borawska, J.; Darewicz, M.; Pliszka, M.; Vegarud, G. E. Antioxidant Properties of Salmon (*Salmo salar* L.) Protein Fraction Hydrolysates Revealed Following Their Ex Vivo Digestion and In Vitro Hydrolysis. *J. Sci. Food Agric.* **2016**, *96*, 2764–2772.

Bougatef, A.; Ravallec, R.; Nedjar-Arroume, N.; Barkia, A.; Guillochon, D.; Nasri, M. Evidence of In Vivo Satietogen Effect and Control of Food Intake of Smooth Hound (*Mustelus mustelus*) Muscle Protein Hydrolysate in Rats. *J. Funct. Food* **2010**, *2*, 10–16.

Burgos-Hernández, A.; Rosas-Burgos, E.; Martínez, M.; Nuncio-Jauregui, P.; Marhuenda, F.; Kačániová, M.; Petrová, J.; Carbonell-Barrachina, A. Bioactive Fractions from Cantabrian Anchovy (*Engraulis encrarischolus*) Viscera. *Food Sci. Technol. Campinas* **2016**, *36*, 426–431.

Caballero, B. *Guide to Nutritional Supplements: Technology & Engineering*; Academic Press: London, **2009**; p 548.

Cai, L.; Wu, X.; Zhang, Y.; Li, X.; Ma, S.; Li, J. Purification and Characterization of Three Antioxidant Peptides from Protein Hydrolysate of Grass Carp (*Ctenopharyngodon idella*) Skin. *J. Funct. Foods* **2015**, *16*, 234–242.

Caruso, G. Fishery Wastes and By-products: A Resource to Be Valorised. *J. Fish Sci. Com.* **2016**, *10*, 12–15.

Caruso, G.; Maricchiolo, G.; Genovese, L.; De Pasquale, F.; Caruso, R.; Denaro, M. G.; Santi Delia, S.; Laganà, P. Comparative Study of Antibacterial and Haemolytic Activities in Sea Bass, European Eel and Blackspot Seabream. *Open Mar. Biol. J.* **2014**, *8*, 10–16.

Cavallini, M.; Gazzola, R.; Metalla, M.; Vaienti, L. The Role of Hyaluronidase in the Treatment of Complications From Hyaluronic Acid Dermal Fillers. *Aesth. Surg. J.* **2013**, *33*, 1167–1174.

Chalamaiah, M.; Hemalatha, R.; Jyothirmayi, T.; Diwan, P. V.; Bhaskarachary, K.; Vajreswari, A.; Kumar, R. R.; Kumar, B. D. Chemical Composition and Immunomodulatory Effects of Enzymatic Protein Hydrolysates from Common Carp (*Cyprinus carpio*) Egg. Nutrition **2015a**, *31*, 388–398.

Chalamaiah, M.; Jyothirmayi, T.; Diwan, P. V.; Kumar, B. D. Antioxidant Activity and Functional Properties of Enzymatic Protein Hydrolysates from Common Carp (*Cyprinus carpio*) Roe (Egg). *J. Food Sci. Technol.* **2015b**, *52*, 5817–5825.

Chan, K. O.; Tong, H. H.; Ng, G. Y. Topical Fish Oil Application Coupling with Therapeutic Ultrasound Improves Tendon Healing. *Ultrasound Med. Biol.* **2016**, *42*, 2983–2989.

Chang, W. T.; Pan, C. Y.; Rajanbabu, V.; Cheng, C. W.; Chen, J. Y. Tilapia (*Oreochromis mossambicus*) Antimicrobial Peptide, Hepcidin 1–5, Shows Antitumor Activity in Cancer Cells. *Peptides* **2011**, *32*, 342–352.

Chee, P. Y.; Mang, M.; Lau, E. S.; Tan, L. T.; He, Y.; Lee, W.; Pusparajah, P.; Chan, K.; Lee, L.; Goh, B. Epinecidin-1, an Antimicrobial Peptide Derived From Grouper (*Epinephelus coioides*): Pharmacological Activities and Applications. *Front. Microbiol.* **2019**. https://doi.org/10.3389/fmicb.2019.02631.

Chen, D. L.; Kini, R. M.; Yuen, R.; Khoo, H. E. Haemolytic Activity of Stonustoxin from Stonefish (*Synanceja horrida*) Venom: Pore Formation and the Role of Cationic Amino Acid Residues. *Biochem. J.* **1997**, *325*, 685–691.

Chen, J.; Chen, Y.; Xia, W.; Xiong, Y. L.; Ye, R.; Wang, H. Grass Carp Peptides Hydrolysed by the Combination of Alcalase and Neutrase: Angiotensin-I Converting Enzyme (ACE) Inhibitory Activity, Antioxidant Activities and Physicochemical Profiles. *Int. J. Food Sci. Technol.* **2016**, *51*, 499–508.

Chen, Y. P.; Liang, C. H.; Wu, H. T.; Pang, H. Y.; Chen, C.; Wang, G. H.; Chan, L. P. Antioxidant and Anti-Inflammatory Capacities of Collagen Peptides from Milkfish (*Chanos chanos*) Scales. *J. Food Sci. Technol.* **2018**, *55*, 2310–2317.

Chen, Y.; Jin, H.; Yang, F.; Jin, S.; Liu, C.; Zhang, L.; Huang, J.; Wang, S.; Yan, Z.; Cai, X.; Zhao, R.; Yu, F.; Yang, Z.; Ding, G.; Tang, Y. Physicochemical, Antioxidant Properties of Giant Croaker (*Nibea japonica*) Swim Bladders Collagen and Wound Healing Evaluation. *Int. J. Biol. Macromol.* **2019**, *138*, 483–491.

Cheung, R. C. F.; Ng, T. B.; Wong, J. H. Marine Peptides: Bioactivities and Applications. *Mar. Drugs* **2015**, *13*, 4006–4043.

Chi, C. F.; Wang, B.; Wang, Y. M.; Zhang, B.; Deng, S. G. Isolation and Characterization of Three Antioxidant Peptides from Protein Hydrolysate of Bluefin Leatherjacket (*Navodon septentrionalis*) Heads. *J. Funct. Foods* **2015a**, *12*, 1–10.

Chi, C.; Wang, B.; Hu, F.; Wang, Y.; Zhang, B.; Deng, S.; Wu, C. Purification and Identification of Three Novel Antioxidant Peptides from Protein Hydrolysate of Bluefin Leatherjacket (*Navodon septentrionalis*) Skin Food Res. Int. **2015b**, *73*, 124–129.

Chi, C. F.; Hu, F. Y.; Wang, B.; Ren, X. J.; Deng, S. G.; Wu, C. W. Purification and Characterization of Three Antioxidant Peptides from Protein Hydrolyzate of Croceine Croaker (*Pseudosciaena crocea*) Muscle. *Food Chem.* **2015c**, *168*, 662–667.

Cho, J.; Kim, Y. Sharks: A Potential Source of Antiangiogenic Factors and Tumor Treatments. *Mar. Biotechnol. N.Y.* **2002**, *4*, 521–525.

Ciarlo, A. S.; Paredi, M. E.; Fraga, A. N. Isolation of Soluble Collagen from Hake Skin (*Merluccius hubbsi*). *J. Aquat. Food Prod. Technol.* **1997**, *6*, 65–77.

Cinq-Mars, C. D.; Li-Chan, E. C. Y. Optimizing Angiotensin I-Converting Enzyme Inhibitory Activity of Pacific Hake (*Merluccius productus*) Fillet Hydrolysate Using Response Surface Methodology and Ultrafiltration. *J. Agric. Food Chem.* **2007**, *55*, 9380–9388.

Cohut, M. *How a Parasitic Fish could Help Us Fight Brain Cancer and Stroke*, **2019**. https://www.medicalnewstoday.com/articles/325211.php#1.

Cole, A. M.; Weis, P.; Diamond, G. Isolation and Characterization of Pleurocidin, an Antimicrobial Peptide in the Skin Secretions of Winter Flounder. *J. Biol. Chem.* **1997**, *272*, 12008–12013.

Conceição, K.; Monteiro-Dos-Santos, J.; Seibert, C., Silva, P.; Marques, E.; Richardson, M.; Lopes-Ferreira, M. *Potamotrygon henlei* Stingray Mucus: Biochemical Features of a Novel Antimicrobial Protein. *Toxicon* **2012**, *60*, 821–829.

Conrad, D.; Gil-Agudelo, D. L.; Ritchie, K. B. Antibiotic-Producing Bacteria Associated with Sharks of St. Helena and Port Royal Sound. **2018**. https://digitalcommons.northgeorgia.edu/gurc/2018/masterschedule/19/ (conference poster abstract).

Cudennec, B.; Fouchereau-Peron, M.; Ferry, F.; Duclos, E.; Ravallec, R. In Vitro and In Vivo Evidence for a Satiating Effect of Fish Protein Hydrolysate Obtained from Blue Whiting (*Micromesistius poutassou*) Muscle. *J. Funct. Food* **2012**, *4*, 271–277.

Darewicz, M.; Borawska, J.; Vegarud, G. E.; Minkiewicz, P.; Iwaniak, A. Angiotensin I-Converting Enzyme (ACE) Inhibitory Activity and ACE Inhibitory Peptides of Salmon (*Salmo salar*) Protein Hydrolysates Obtained by Human and Porcine Gastrointestinal Enzymes. *Int. J. Mol. Sci.* **2014**, *15*, 14077–14101.

da Rocha, M.; Loiko, M. R.; Tondo, E. C.; Prenticea, C. Physical, Mechanical and Antimicrobial Properties of Argentine Anchovy (*Engraulis anchoita*) Protein Films Incorporated with Organic Acids. Food Hydrocoll*oids* **2014**, *37*, 213–220.

De, B.; Deb, S. R.; Chakraborty, S.; Namasudra, U.; Pal, M. R.; Choudhury, R.; Goswami, B. B.; Datta, S. P.; Sen, S.; Chakraborty, R. Antibacterial and Antidiabetic Evaluation of Bile Content of *Catla catla* & *Labeo rohita*. *Cen. Eur. J. Exp. Biol.* **2012**, *1*, 107–112.

de Amorim, R. G. O; Deschamps, F. C; Pessatti, M. L. Protein Hydrolysate Waste of Whitemouth Croaker (*Micropogonias furnieri*) as a Way of Adding Value to Fish and Reducing the Environmental Liabilities of the Fishing Industry. *Lat. Am. J. Aquat. Res.* **2016**, *44*, 967–974.

Deivasigamani, B.; Subramanian, V.; Sundaresan, A. Identification of Protein from Muscle Tissue of Marine Finfish. *Int. J. Curr. Microbiol. Appl. Sci.* **2017**, *6*, 159–167.

Deo, A. D.; Venkateshvaran, K.; Devaraj, M. Bioactivity of Epidermal Secretion of Two Marine Catfish from Mumbai. *J. Indian Fish Assoc.* **2008**, *5*, 97–111.

Donati, M., Francesco, A. D.; Paolo, M. D.; Fiani, N.; Benincasa, M.; Gennaro, R.; Nardini, P.; Foschi, C.; Cevenini, R. Activity of Cathelicidin Peptides against *Simkania negevensis*. *Int. J. Pept.* **2011**, 3, Article ID 708710.

Dong, Y.; Sheng, G.; Fu, J.; Wen, K. Chemical Characterization and Anti-Anaemia Activity of Fish Protein Hydrolysate from *Saurida elongata*. *J. Sci. Food. Agric.* **2005**, *85*, 2033–2039.

Eastoe, J. E.; Leach, A. A. Chemical Constitution of Gelatine. In *The Science and Technology of Gelatine*; Ward, A. G., Courts, A., Eds.; Academic Press: London, **1977**; pp 73–107.

Ekanayake, P. M.; Park, G. A.; Lee, Y. D.; Kim, S. J.; Jeong, S. C.; Lee, J. Antioxidant Potential of Eel (*Anguilla japonica* and *Conger myriaster*) Flesh and Skin. *J. Food Lipids* **2005**, *22*, 34–47.

Elavarasan, K.; Naveen Kumar, V. N.; Shamasundar, B. A. Antioxidant and Functional Properties of Fish Protein Hydrolysates from Fresh Water Carp (*Catla catla*) as Influenced by the Nature of Enzyme. *J. Food Process Preserv.* **2014**, *38*, 1207–1214.

Enari, H.; Takahashi, Y.; Kawarasaki, M.; Tada, M.; Tatsuta, K. Identification of Angiotensin I-converting Enzyme Inhibitory Peptides Derived from Salmon Muscle and their Antihypertensive Effect. *Fish Sci.* **2008**, *74*, 911–920.

Falkenberg, S. S.; Mikalsen, S. O.; Joensen, H.; Stagsted, J.; Nielsen, H. H. Extraction and Characterization of Candidate Bioactive Compounds in Different Tissues from Salmon (*Salmo salar*). *Int. J. Appl. Res. Nat. Prod.* **2014**, *7*, 11–25.

Ferdosh, S.; Sarker, M. Z. I.; Rahman, N. N. N. A.; Akand, M. J. H.; Ghafoor, K.; Awang, M. B.; Kadir, M. O. A. Supercritical Carbon Dioxide Extraction of Oil from *Thunnus tonggol* Head by Optimization of Process Parameters using Response Surface Methodology. *Korean J. Chem. Eng.* **2013**, *30*, 1466–1472.

Fu, Y.; Zhao, X. H. *In Vitro* Responses of hFOB1.19 Cells towards Chum Salmon (*Oncorhynchus keta*) Skin Gelatin Hydrolysates in Cell Proliferation, Cycle Progression and Apoptosis. *J. Funct. Foods* **2013**, *5*, 279–288.

Fu, W.; Chen, C.; Zeng, H.; Lin, J.; Zhang, Y.; Hu, J.; Zheng, B. Novel Angiotensin-Converting Enzyme Inhibitory Peptides Derived from *Trichiurus lepturus* Myosin: Molecular Docking and Surface Plasmon Resonance Study. *LWT—Food Sci. Technol.* **2019**, *110*, 54–63.

Fujita, H.; Yoshikawa, M. LKPNM: A Prodrug-Type ACE-Inhibitory Peptide Derived from Fish Protein. *Immunopharmacology* **1999**, *44*, 123–127.

Galla, N. R.; Pamidighantam, P. R.; Akula, S.; Karakala, B. Functional Properties and In Vitro Antioxidant Activity of Roe Protein Hydrolysates of *Channa striatus* and *Labeo rohita*. *Food Chem.* **2012**, *135*, 1479–1484.

Galvez, R. P.; Berge, J., Eds. *Utilization of Fish Waste*; CRC Press: Boca Raton, **2013**.

Gamarro, E. G.; Orawattanamateekul, W.; Sentina, J.; Gopal, T. K. S. By-Products of Tuna Processing. *Globefish Research Programme*, 2013, 112. http://www.fao.org/3/a-bb215e.pdf.

Ganesh, R. J.; Nazeer, R. A.; Kumar, N. S. S. Purification and Identification of Antioxidant Peptide from Black Pomfret, *Parastromateus niger* (Bloch, 1975) Viscera Protein Hydrolysate. Food Sci. Biotechnol. **2011**, *20*, 1087.

García-Moreno, P. J.; Pérez-Gálvez, R.; Espejo-Carpio, F. J.; Ruiz-Quesada, C.; Pérez-Morilla, A. I.; Martínez-Agustín, O.; Guadix, A.; Guadix, E. M. Functional, Bioactive and Antigenicity Properties of Blue Whiting Protein Hydrolysates: Effect of Enzymatic Treatment and Degree of Hydrolysis. *J. Sci. Food Agric.* **2017**, *97*, 299–308.

Gevaert, B.; Veryser, L.; Verbeke, F.; Wynendaele, E.; De Spiegeleer, C. Fish Hydrolysates: A Regulatory Perspective of Bioactive Peptides. *Protein Peptide Lett.* **2016**, *23*, 1–9.

Ghaly, A. E.; Ramakrishnan, V. V.; Brooks, M. S.; Budge, S. M.; Dave, D. Fish Processing Wastes as a Potential Source of Proteins, Amino Acids and Oils: A Critical Review. *J. Microb. Biochem. Technol.* **2013**, *5*, 107–129.

Ghelichi, S.; Shabanpour, B.; Pourashouri, P.; Hajfathalian, M.; Jacobsen, C. Extraction of Unsaturated Fatty Acid-Rich Oil from Common Carp (*Cyprinus carpio*) Roe and Production of Defatted Roe Hydrolysates with Functional, Antioxidant, and Antibacterial Properties. *J. Sci. Food Agric.* **2018**, *98*, 1407–1415.

Giri, A.; Nasu, M.; Ohshima, T. Bioactive Properties of Japanese Fermented Fish Paste, Fish Miso, Using Koji Inoculated with *Aspergillus oryzae*. *Int. J. Food Sci. Nutr.* **2012**, *1*, 13–22.

Gómez-Guillén, M. C.; López-Caballero, M. E.; de Lacey, A. A. A. L.; Giménez, B.; Montero, P. Antioxidant and Antimicrobial Peptide Fractions from Squid and Tuna Skin Gelatin. *Sea By-products as Real Material: New Ways of Application*; **2010**; pp 89–115.

González, R. P.; Leyva, A.; Moraes, M. O. Shark Cartilage as Source of Antiangiogenic Compounds: from Basic to Clinical Research. *Biol. Pharm. Bull.* **2001**, *24*, 1097–1101.

Granito, R. N.; Renno, A. C. M.; Yamamura, H.; de Almeida, M. C.; Ruiz, P. L. M.; Ribeiro, D. A. Hydroxyapatite from Fish for Bone Tissue Engineering: A Promising Approach. *Int. J. Mol. Cell. Med.* **2018,** *7,* 80–90.

Greenberg, J. A.; Bell, S. J.; Ausdal, W. V. Omega-3 Fatty Acid Supplementation during Pregnancy. *Rev. Obstet. Gynecol.* **2008,** *1,* 162–169.

Guérard, F.; Sellos, D.; Gal, Y. L. Fish and Shellfish Upgrading, Traceability. *Adv. Biochem. Eng./Biotechnol.* **2005,** *96,* 127–163.

Hamaguchi, P. Y.; Bergsson, A. B.; Halldorsdottir, S. M.; Thorkelsson, G.; Kristinsson, H. G.; Johannsson, R. Bioactivity of Saithe (*Pollachius virens* L.) Protein Hydrolysates. In *5th World Fisheries Congress*, Yokohama, Japan, **2008.**

Han, S. Preparation of Collage and Collagen Peptides from Bluefin Tuna Skin (Bone and Scale) and Their Action on Stressed HepG2 Cell. Education Achievements Report, Graduate School of Agriculture, Kinki University, n.d.

Han, Y.; Byun, S. H.; Park, Y. H.; Kim, S. B. Bioactive Properties of Enzymatic Hydrolysates from Abdominal Skin Gelatin of Yellowfin Tuna (*Thunnus albacares*). *Food Sci. Technol.* **2015,** *50,* 1996–2003.

He, L.; Lan, W.; Wang, Y.; Ahmed, S.; Liu, Y. Extraction and Characterization of Self-Assembled Collagen Isolated from Grass Carp and Crucian Carp. *Foods* **2019,** *8,* 396.

Hernández-Ledesma, B.; Herrero, M., Eds. *Bioactive Compounds from Marine Foods: Plant and Animal Sources*; Technology and Engineering; John Wiley & Sons: Hoboken, NJ, **2013**; p 464.

Hosomi, R.; Fukunaga, K.; Arai, H.; Kanda, S.; Nishiyama, T.; Yoshida, M. Effect of Combination of Dietary Fish Protein and Fish Oil on Lipid Metabolism in Rats. *J. Food Sci. Technol.* **2013,** *50,* 266–274.

Hosseini, S. V.; Sobhanardakani, S; Tahergorabi, R; Delfieh, P. Selected Heavy Metals Analysis of Persian Sturgeon's (*Acipenser persicus*) Caviar from Southern Caspian Sea. *Biol. Trace Elem. Res.* **2013,** *154,* 357–362.

Hsieh, C. H.; Shiau, C. Y.; Su, Y. C.; Liu, Y. H.; Huang, Y. R. Isolation and Characterization of Collagens from the Skin of Giant Grouper (*Epinephelus lanceolatus*). *J. Aquat. Food Prod. Technol.* **2014,** *25,* 93–104.

Hsu, K. C. Purification of Antioxidative Peptides Prepared from Enzymatic Hydrolysates of Tuna Dark Muscle By-product. *Food Chem.* **2010,** *122,* 42–48.

Hsu, K. C.; Li-Chan, E. C. Y.; Jao, C. L. Antiproliferative Activity of Peptides Prepared from Enzymatic Hydrolysates of Tuna Dark Muscle on Human Breast Cancer Line MCF-7. *Food Chem.* **2011,** *126,* 617–622.

Hu, Z.; Yang, P.; Zhou, C.; Li, S; Hong, P. Marine Collagen Peptides from the Skin of Nile Tilapia (*Oreochromis niloticus*): Characterization and Wound Healing Evaluation. *Mar. Drugs* **2017,** *15,* 102.

Huang, P. H.; Chen, J. Y.; Kuo, C. M. Three Different Hepcidins from Tilapia, *Oreochromis mossambicus*: Analysis of Their Expressions and Biological Functions. *Mol. Immunol.* **2007,** *44,* 1922–1934.

Huang, Y. R.; Shiau, C. Y.; Chen, H. H.; Huang, B. C. Isolation and Characterization of Acid and Pepsin-Solubilized Collagens from the Skin of Balloon Fish (*Diodon holocanthus*). *Food Hydrocolloids* **2011,** *25,* 1507–1513.

Huang, T. H.; Wang, P. W.; Yang, S. C.; Chou, W. L.; Fang, J. Y. Cosmetic and Therapeutic Applications of Fish Oil's Fatty Acids on the Skin. *Mar. Drugs* **2018,** *16,* pii: E256.

Hukmi, N. M. M.; Sarbon, N. M. Isolation and Characterization of Acid Soluble Collagen (ASC) and Pepsin Soluble Collagen (PSC) Extracted from Silver Catfish (*Pangasius* sp.) Skin. *Int. Food Res. J.* **2018**, *25*, 1785–1791.

Ichimura, T.; Hu, J.; Aita, D. Q.; Maruyama, S. Angiotensin I-Converting Enzyme Inhibitory Activity and Insulin Secretion Stimulative Activity of Fermented Fish Sauce. *J. Biosci. Bioeng.* **2003**, *96*, 496–499.

Ikeda, M. Amino Acid Production Processes. In *Advances in Biochemical Engineering/Biotechnology*; Scheper, T., Faurie, R., Thommel, J., Eds.; Springer: Berlin-Heidelberg, New York, USA, **2003**; Vol 79, pp 1–35.

Indumathi, S. M.; Manigandan, V.; Khora, S. S. Antimicrobial and Larvicidal Activities of the Tissue Extracts of Oblong Blowfish (*Takifugu oblongus*) from South-East Coast of India. *Int. J. Toxicol. Pharmacol. Res.* **2016**, *8*, 312–319.

Irwandi, J.; Faridayanti, S.; Mohamed, E. S. M.; Hamzah, M. S.; Torla, H. H.; Che Man, Y. B. Extraction and Characterization of Gelatin from Different Marine Fish Species in Malaysia. *Int. Food Res. J.* **2009**, *16*, 381–389.

Ishimura, G.; Bailey, M. The Market Value of Freshness: Observations from the Swordfish and Blue Shark Longline Fishery. *Fish Sci.* **2013**, *79*, 547–553.

Jais, A. M. M.; Zakaria, Z. A.; Luo, A.; Song, Y. X. Antifungal Activity of *Channa striatus* (Haruan) Crude Extracts. *Int. J. Trop. Med.* **2008**, *3*, 43–48.

Jal, S.; Priya, K. M.; Mandal, N.; Khora, S. S. Bioactive Potential of Puffer Fish *Arothron stellatus* Collected from South East Coast of India. *Int. J. Drug Dev. Res.* **2014**, *6*, 102–108.

Jao, C.; Ko, W. 1,1-Diphenyl-2-picrylhydrazyl (DPPH) Radical Scavenging by Protein Hydrolyzates from Tuna Cooking Juice. *Fish Sci.* **2002**, *68*, 430–435.

Je, J.; Park, P.; Kwon, J. Y.; Kim, S. A Novel Angiotensin I Converting Enzyme Inhibitory Peptide from Alaska Pollack (*Theragra chalcogramma*) Frame Protein Hydrolysate. *J. Agric. Food Chem.* **2004**, *52*, 7842–7845.

Je, J.; Park, P.; Kim, S. Antioxidant Activity of a Peptide Isolated from Alaska Pollack (*Theragra chalcogramma*) Frame Protein Hydrolysate. *Food Res. Int.* **2005**, *38*, 45–50.

Je, J. Y.; Qian, Z. J.; Lee, S. H.; Byun, H. G.; Kim, S. K. Purification and Antioxidant Properties of Bigeye Tuna (*Thunnus obesus*) Dark Muscle Peptide on Free Radical-Mediated Oxidative Systems. *J. Med. Food* **2008**, *11*, 629–637.

Jeevitha, K.; Priya, K. M.; Khora, S. S. Antioxidant activity of Fish Protein Hydrolysates from *Sardinella longiceps*. *Int. J. Drug Dev. Res.* **2014**, *6*, 137–145.

Jeevithan, E.; Bao, B.; Bu, Y.; Zhou, Y.; Zhao, Q.; Wu, W. Type II Collagen and Gelatin from Silvertip Shark (*Carcharhinus albimarginatus*) Cartilage: Isolation, Purification, Physicochemical and Antioxidant Properties. *Mar. Drugs* **2014**, *12*, 3852–3873.

Jemil, I.; Abdelhedi, O.; Nasri, R.; Mora, L.; Jridi, M.; Aristoy, M.; Toldrá, F.; Nasri, M. Novel Bioactive Peptides from Enzymatic Hydrolysate of Sardinelle (*Sardinella aurita*) Muscle Proteins Hydrolysed by *Bacillus subtilis* A26 Proteases. *Food Res. Int.* **2017**, 121–133.

Jensen, K. N.; Jacobsen, C.; Nielsen, H. H. Fatty Acid Composition of Herring (*Clupea harengus* L.): Influence of Time and Place of Catch on n-3 PUFA Content. *J Sci. Food Agric.* **2007**, *87*, 710–718.

Jia, J.; Zhou, Y.; Lu, J.; Chen, A.; Li, Y.; Zheng, G. Enzymatic Hydrolysis of Alaska Pollack (*Theragra chalcogramma*) Skin and Antioxidant Activity of the Resulting Hydrolysate. *J. Sci. Food Agric.* **2010**, *90*, 635–640.

Jiang, H.; Tong, T.; Sun, J.; Xu, Y.; Zhao, Z.; Liao, D. Purification and Characterization of Antioxidative Peptides from Round Scad (*Decapterus maruadsi*) Muscle Protein Hydrolysate. *Food Chem.* **2014**, *154*, 158–163.

John, S. T.; Velmurugan, S.; Nagaraj, D. S.; Lisa, A. Antimicrobial Activity of Striped Eel Cat Fish *Plotosus lineatus* against Human Pathogens. *World J. Pharm. Sci.* **2015**, *3*, 1134–1137.

Jridia, M.; Abdelhedi, O.; Zouariac, N.; Fakhfakh, N.; Nasri, M. Development and Characterization of Grey Triggerfish Gelatin/Agar Bilayer and Blend Films Containing Vine Leaves Bioactive Compounds. *Food Hydrocolloids* **2019**, *89*, 370–378.

Jung, W.; Lee, B.; Kim, S. Fish-Bone Peptide Increases Calcium Solubility and Bioavailability in Ovariectomised Rats. *Br. J. Nutr.* **2006a**, *95*, 124–128.

Jung, W. K.; Mendis, E.; Je, J. Y.; Park, P. J.; Son, B. W.; Kim, H. C.; Choi, Y. K.; Kim, S. Angiotensin I-Converting Enzyme Inhibitory Peptide from Yellowfin Sole (*Limanda aspera*) Frame Protein and Its Antihypertensive Effect in Spontaneously Hypertensive Rats. *Food Chem.* **2006b**, *94*, 26–32.

Junianto, I.; Rizal, A. Characteristics of Physical–Chemical Properties of Collagen Extracted from the Skin of Bonylip Barb Fish (*Osteochilus vittatus*). *World App. Sci. J.* **2018**, *36*, 78–84.

Kaleshkumar, K.; Rajaram, R.; Gayathri, N.; Sivasudha, T.; Arun, G.; Archunan, G; Gulyás, B.; Padmanabhan, P. Muscle Extract of *Arothron immaculatus* Regulates the Blood Glucose Level and the Antioxidant System in High-Fat Diet and Streptozotocin Induced Diabetic Rats. *Bioorg. Chem.* **2019**, *90*, 103072.

Kalidasan, K.; Ravi, V.; Sahu, S. K.; Maheshwaran, M. L.; Kandasamy, K. Antimicrobial and Anticoagulant Activities of the Spine of Stingray *Himantura imbricata*. *J. Coast Life Med.* **2014**, *2*, 89–93.

Karim, A. A.; Bhat, R. Gelatin Alternatives for the Food Industry: Recent Developments, Challenges and Prospects. *Trends Food Sci. Technol.* **2008**, *19*, 644–656.

Karim, A. A.; Bhat, R. Fish Gelatine: Properties, Challenges, and Prospects as an Alternative to Mammalian Gelatines. *Food Hydrocolloid* **2009**, *23*, 563–576.

Karimzadeh, K. Antihypertensive and Anticoagulant Properties of Glycosaminoglycans Extracted from the Sturgeon (*Acipenser persicus*) Cartilage. *Curr. Issues Pharm. Med. Sci.* **2018**, *31*, 163–169.

Karmakar, S.; Dasgupta, S. C.; Gomes, A. Pharmacological and Haematological Study of Shol Fish (*Channa striatus*) Skin Extract on Experimental Animal. *Indian J. Exp. Biol.* **2002**, *40*, 115–118.

Kato, K.; Nakagawa, H.; Shinohara, M.; Ohura, K. Purification of a Novel Lectin from the Dorsal Spines of the Stonefish, *Synanceia verrucosa*. *J. Osaka Dent. Univ.* **2016**, *50*, 55–61.

Ke, F.; Wang, Y.; Yang, C.; Xu, C. Molecular Cloning and Antibacterial Activity of Hepcidin from Chinese Rare Minnow (*Gobiocypris rarus*). *Electron. J. Biotechn.* **2015**, *18*, 169–174.

Khaled, H. B.; Ktari, N.; Ghorbel-Bellaaj, O.; Jridi, M.; Lassoued, I.; Nasri, M. Composition, Functional Properties and In Vitro Antioxidant Activity of Protein Hydrolysates Prepared from Sardinelle (*Sardinella aurita*) Muscle. *J. Food Sci. Technol.* **2014**, *51*, 622–633.

Khantaphant, S.; Benjakul, S.; Kishimura, H. Antioxidative and ACE Inhibitory Activities of Protein Hydrolysates from the Muscle of Brownstripe Red Snapper Prepared Using Pyloric Caeca and Commercial Proteases. *Process Biochem.* **2011**, *46*, 318–327.

Khiari, Z. Functional and Bioactive Components from Mackerel (*Scomber scombrus*) and Blue Whiting (*Micromesistius poutassou*) Processing Waste. Ph.D. Thesis, Dublin Institute of Technology, **2010**.

Khoo, H. E.; Yuen, R.; Poh, C. H.; Tan, C. H. Biological Activities of *Synanceja horrida* (stonefish) Venom. *Mol. Genet. Genom. Med.* **1992**, *1*, 54–60.

Khora, S. S. Marine Fish-Derived Bioactive Peptides and Proteins for Human Therapeutics. *Int. J. Pharm. Pharm. Sci.* **2013**, *5*, 31–37.

Kim, S. Peptide-Derived from Seahorse Exerts a Protective Effect against Cholinergic Neuronal Death in In Vitro Model of Alzheimer's Disease. *Ratih Pangestuti Proc. Chem.* **2015**, *14*, 343–352.

Kim, S. R.; Byun, H. G. The Novel Angiotensin I Converting Enzyme Inhibitory Peptide from Rainbow Trout Muscle Hydrolysate. *Fish Aquat. Sci.* **2012**, *15*, 183–190.

Kim, S. K.; Karadeniz, F. Biological Importance and Applications of Squalene and Squalane. *Adv. Food Nutr. Res.* **2012**, *65*, 223–233.

Kim, J.; Cho, M.; Heu, M. Preparation of Calcium Powder from Cooking Skipjack Tuna Bone and Its Characteristics. *J. Korean Fish. Soc.* **2000**, *33*, 158–163.

Kim, S.; Kim, Y.; Byun, H.; Nam, K.; Joo, D.; Shahidi, F. Isolation and Characterization of Antioxidative Peptides from Gelatin Hydrolysate of Alaska Pollack, Skin. *J. Agric. Food Chem.* **2001**, *49*, 1984–1989.

Kim, H. S.; Choi, E. O.; Kim, M. D.; Choi, Y. H.; Kim, B. W.; Kim, S. Y.; Hwang, H. J. Effect of Calcium Extracted from Salted Anchovy (*Engraulis japonicus*) on Calcium Metabolism of the Rat. *J. Korean Soc. Food Sci. Nutr.* **2013**, *42*, 182–187.

Kitani, Y.; Kikuchi, N.; Zhang, G.; Ishizaki, S.; Shimakura, K.; Shiomi, K.; Nagashima, Y. Antibacterial Action of L-Amino Acid Oxidase from the Skin Mucus of Rockfish *Sebastes schlegelii*. *Comp. Biochem. Physiol. B: Biochem. Mol. Biol.* **2007**, *149*, 394–400.

Kittiphattanabawon, P.; Benjakul, S.; Visessanguan, W.; Nagai, T.; Tanaka, M. Characterisation of Acid-Soluble Collagen from Skin and Bone of Bigeye Snapper (*Priacanthus tayenus*). *Food Chem.* **2005**, *89*, 363–372.

Kittiphattanabawon, P.; Benjakul, S.; Visessanguan, W.; Shahidi, F. Isolation and Properties of Acid- and Pepsin-Soluble Collagen from the Skin of Blacktip Shark (*Carcharhinus limbatus*). *Eur. Food Res. Technol.* **2010a**, *230*, 475–483.

Kittiphattanabawon, P.; Benjakul, S.; Visessanguan, W.; Kishimura, H.; Shahidi, F. Isolation and Characterisation of Collagen from the Skin of Brownbanded Bamboo Shark (*Chiloscyllium punctatum*). *Food Chem.* **2010b**, *119*, 1519–1526.

Kittiphattanabawon, P.; Benjakul, S.; Visessanguan, W.; Shahidi, F. Isolation and Characterization of Collagen from the Cartilages of Brown-Banded Bamboo Shark (*Chiloscyllium punctatum*) and Blacktip Shark (*Carcharhinus limbatus*). LWT—Food Sci. Technol. **2010c**, *43*, 792–800.

Ko, J. Y.; Lee, J. H.; Samarakoon, K.; Kim, J. S.; Jeon, Y. J. Purification and Determination of Two Novel Antioxidant Peptides from Flounder Fish (*Paralichthys olivaceus*) Using Digestive Proteases. *Food Chem. Toxicol.* **2013**, *52*, 113–120.

Kong, Z.; Gau, S.; Feng, K.; Chen, T. *Animal Cell Technology: Basic & Applied Aspects: Volume 12*: *Animal Cell Technology for Creation of New Era*. In Japanese Association for Animal Cell Technology Meeting, Springer Science & Business Media: Berlin, **2002**; pp 449–454.

Korczek, K.; Tkaczewska, J.; Migdał, W. Antioxidant and Antihypertensive Protein Hydrolysates in Fish Products—A Review. *Czech. J. Food Sci.* **2018,** *36,* 195–207.

Kristinsson, H. G. *Antioxidants and Functional Components in Aquatic Foods*; John Wiley & Sons: Hoboken, NJ, **2014**; p 344.

Ktari, N.; Jridi, M.; Bkhairia, I.; Sayari, N. Functionalities and Antioxidant Properties of Protein Hydrolysates from Muscle of Zebra Blenny (*Salaria basilisca*) Obtained with Different Crude Protease Extracts. *Food Res. Int.* **2012,** *49,* 747–756.

Ktari, N.; Mnafgui, K.; Nasri, R.; Hamden, K.; Bkhairia, I.; Ben Hadj, A. B.; Boudaouara, T.; Elfeki, A.; Nasri, M. Hypoglycemic and Hypolipidemic Effects of Protein Hydrolysates from Zebra Blenny (*Salaria basilisca*) in Alloxan-Induced Diabetic Rats. *Food Funct.* **2013,** *4,* 1691–1699.

Ktari, N.; Nasri, R.; Mnafgui, K.; Hamden, K.; Belguith, O.; Boudaouara, T.; El Feki, A.; Nasri, M.; Antioxidative and ACE Inhibitory Activities of Protein Hydrolysates from Zebra Blenny (*Salaria basilisca*) in Alloxan-Induced Diabetic Rats (English). *Process Biochem.* **2014,** *49,* 890–897.

Ktari, N.; Belguith-Hadriche, O.; Amara, I. B.; Hadj, A. B.; Turki, M.; Makni-Ayedi, F; Boudaouara, T.; El Feki, A.; Boualga, A.; Salah, R. B.; Nasri, M. Cholesterol Regulatory Effects and Antioxidant Activities of Protein Hydrolysates from Zebra Blenny (*Salaria basilisca*) in Cholesterol-Fed Rats. *Food Funct.* **2015,** *6,* 2273–2282.

Kumar, N. S. S.; Nazeer, R. A. Characterization of Acid and Pepsin Soluble Collagen from the Skin of Horse Mackerels (*Magalaspis cordyla*) and Croaker (*Otolithes ruber*). *Int. J. Food Prop.* **2013,** *16,* 613–621.

Kumar, N.; Nazeer, R.; Jaiganesh, R. Purification and Biochemical Characterization of Antioxidant Peptide from Horse Mackerel (*Magalaspis cordyla*) Viscera Protein. *Peptides* **2011,** *32,* 1496–1501.

Kumar, N. S. S.; Nazeer, R. A.; Jaiganesh, R. Wound Healing Properties of Collagen from the Bone of Two Marine Fishes. *Int. J. Pept. Res. Ther.* **2012,** *18,* 185–192.

Kumaravel, K.; Ravichandran, S.; Balasubramanian, T.; Subramanian, K. S.; Bhat, B. A. Antimicrobial Effect of Five Seahorse Species From Indian Coast. *Br. J. Pharmacol. Toxicol.* **2010,** *1,* 62–66.

Kumaravel, K. S.; Joseph, F. R. S; Manikodi, D.; Doimi, M. In Vitro Antimicrobial Activity of Tissue Extracts of Puffer fish *Arothron immaculatus* against Clinical Pathogens. *Chin. J. Nat. Med.* **2011,** *9,* 446–449.

Kurtovic, I.; Marshall, S. N.; Zhao, X.; Simpson, B. K. Lipases from Mammals and Fishes. *Rev. Fish. Sci.* **2009,** *17,* 18–40.

Lamas, D. L.; Yeannes, M. I.; Massa, A. E. Alkaline Trypsin from the Viscera and Heads of *Engraulis anchoita*: Partial Purification and Characterization. *Biotechnologia* **2017,** *98,* 103–112.

Lassoued, I.; Mora, L.; Barkia, A.; Aristoy, M. C.; Nasri, M.; Toldrá, F. Bioactive Peptides Identified in Thornback Ray Skin's Gelatin Hydrolysates by Proteases from *Bacillus subtilis* and *Bacillus amyloliquefaciens*. *J. Proteomics* **2015,** *128,* 8–17.

Le, T.; Maki, H.; Okazaki, E.; Osako, K.; Takahashi, K. Influence of Various Phenolic Compounds on Properties of Gelatin Film Prepared from Horse Mackerel *Trachurus japonicus* Scales. *J. Food Sci.* **2018,** *83,* 1888–1895.

Lee, J. K.; Byun, H. Characterization of Antioxidative Peptide Purified from Black Eelpout (*Lycodes diapterus*) Hydrolysate. *Fish Aquat. Sci.* **2019,** *22,* 22, 7p.

Lee, S. H.; Qian, Z. J.; Kim, S. K. A Novel Angiotensin I Converting Enzyme Inhibitory Peptide from Tuna Frame Protein Hydrolysate and Its Antihypertensive Effect in Spontaneously Hypertensive Rats. *Food Chem.* **2010**, *118*, 96–102.

Lee, W.; Jeon, J.; Byun, H. Characterization of a Novel Antioxidative Peptide from the Sand Eel *Hypoptychus dybowskii*. *Process Biochem.* **2011**, *46*, 1207–1211.

Lee, J. K.; Jeon, J. K.; Byun, H. G. Antihypertensive Effect of Novel Angiotensin I Converting Enzyme Inhibitory Peptide from Chum Salmon (*Oncorhynchus keta*) Skin in Spontaneously Hypertensive Rats. *J. Funct. Foods* **2014**, *7*, 381–389.

Lee, J. K.; Li-Chan, E. C. Y.; Byun, H. G. Characterization of β-Secretase Inhibitory Peptide Purified from Skate Skin Protein Hydrolysate. *Eur. Food Res. Technol.* **2015**, *240*, 129–136.

Levinton, Zh. B; Rogovaia, A. B; Polishchuk, L. R.; Tikhomirova, L. D.; Zvenigorodskaia, I. D. New Type of Commercial Fish, the Triggerfish (*Balistes capriscus*), in Human Nutrition. *Vopr. Pitan.* **1981**, *6*, 43–45.

Li, D.; Williams, J.; Pietras, R. Squalamine and Cisplain Block Angiogenesis and Growth of Human Ovarian Cancer Cells with or without HER-2 Gene Overexpression. *Oncogene* **2002**, *21*, 2805–2814.

Li, Z. R.; Wang, B.; Chi, C. F.; Zhang, Q. H.; Gong, Y. D.; Tang, J. J.; Luo, H. Y.; Ding, G. Isolation and Characterization of Acid Soluble Collagens and Pepsin Soluble Collagens from the Skin and Bone of Spanish Mackerel (*Scomberomorous niphonius*). *Food Hydrocolloids* **2013**, *31*, 103–113.

Li, L.; Zhao, Y.; Yu He, Y.; Chi, C.; Wang, B. Physicochemical and Antioxidant Properties of Acid and Pepsin-Soluble Collagens from the Scales of Miiuy Croaker (*Miichthys miiuy*). *Mar. Drugs* **2018**, *16*, 394.

Lim, J.; Lee, Y.; Sulaiman, A.; Bilung, L. M.; Chong, Y. L. Antibacterial Activity of the Epidermal Mucus of *Barbodes everetti*. *Trends Undergrad. Res.* **2018**, *1*, 40–44.

Lima, M. M.; Vanier, N. L.; Dias, A. R. G.; Zavareze, E.; Prentice, C.; Moreira, A. S. Whitemouth Croaker (*Micropogonias furnieri*) Protein Hydrolysates: Chemical Composition, Molecular Mass Distribution, Antioxidant Activity and Amino Acid Profile. *Int. Food Res. J.* **2019**, *26*, 247–254.

Liu, D.; Liang, L.; Regenstein, J. M.; Peng, Z. Extraction and Characterisation of Pepsin-Solubilised Collagen from Fins, Scales, Skins, Bones, and Swim Bladders of Bighead Carp (*Hypophthalmichthys nobilis*). *Food Chem.* **2012**, *133*, 1441–1448.

Liu, D.; Nikoo, M.; Boran, G.; Zhou, P.; Regenstein, J. M. Collagen and Gelatin. *Ann. Rev. Food Sci. Agric.* **2015**, *6*, 527–557.

Lu, X. J.; Chen, J.; Chen, M. Z.; Lü, J. N.; Shi, Y. H.; Li, H. Y. Hydrolysates of Swim Bladder Collagen from Miiuy Croaker, *Miichthys miiuy*, Enhances Learning and Memory in Mice. *Curr. Top. Nutraceut.* **2010**, *8*, 149–156.

Luer, C. A. Novel Compounds from Shark and Stingray Epidermal Mucus with Antimicrobial Activity against Wound Infection Pathogens. Annual Report. **2012**. https://apps.dtic.mil/dtic/tr/fulltext/u2/a559624.pdf.

Luer, C. A.; Walsh, C. J. Potential Human Health Applications from Marine Biomedical Research with Elasmobranch Fishes. *Fishes* **2018**, *3*, 47.

Magalhaes, M. R.; da Silva, N. J. Jr.; Ulhoa, C. J. A Hyaluronidase from *Potamotrygon motoro* (Freshwater Stingrays) Venom: Isolation and Characterization. *Toxicon* **2008**, *51*, 1060–1067.

Mahadevan, G.; Mohan, K.; Vinoth, J.; Ravi, V. Biotic Potential of Mucus Extracts of Giant Mudskipper *Periophthalmodon schlosseri* (Pallas, 1770) from Pichavaram, Southeast Coast of India. *J. Basic Appl. Zool.* **2019**, *80*, 13.

Mahboob, S.; Haider, S.; Sultana, S.; Al- Ghanim, K. A.; Al-Misned, F.; Al-Balawi, H. F. A.; Ahmad, Z. Isolation and Characterisation of Collagen from the Waste Material of Two Important Freshwater Fish Species. *J. Anim. Plant Sci.* **2014**, *24*, 1802–1810.

Mahboob, S. Isolation and Characterization of Collagen from Fish Waste Material—Skin, Scales, and Fins of *Catla catla* and *Cirrhinus mrigala*. *J. Food Sci. Technol.* **2015**, *52*, 4296–4305.

Malde, M. K.; Graff, I. E.; Siljander-Rasi, H.; Venäläinen, E.; Julshamn, K.; Pedersen, J. I.; Valaja, J. Fish Bones—A Highly Available Calcium Source for Growing Pigs. *J. Anim. Physiol. Anim. Nutr.* **2010**, *94*, e66–e76.

Mathew, S.; Hassan, F. Distribution of Collagen in the Muscle Tissue of Commercially Important Fishes. *J. Food Sci. Technol.* **1996**, *33*, 121–123.

Matmaroh, K.; Benjakul, S.; Prodpran, T.; Encarnacion, A.; Kishimura, H. Characteristics of Acid Soluble Collagen and Pepsin Soluble Collagen from Scale of Spotted Golden Goatfish (*Parupeneus heptacanthus*). *Food Chem.* **2011**, *129*, 1179–1186.

Mendis, E.; Rajapakse, N; Kim, S. K. Antioxidant Properties of a Radical-Scavenging Peptide Purified from Enzymatically Prepared Fish Skin Gelatin Hydrolysate. *J. Agric. Food Chem.* **2005**, *53*, 581–587.

Messina, C. M.; Renda, G.; Barbera, L. L.; Santulli, A. By-products of Farmed European Sea Bass (*Dicentrarchus labrax* L.) as a Potential Source of n-3 PUFA. *Biologia* **2013**, *68*, 288–293.

Migliolo, L.; Felício, M. R.; Cardoso, M. H.; Silva, O. N.; Xavier, M. A.; Nolasco, D. O.; de Oliveira, A. S.; Roca-Subira, I.; Estape, J. V.; Teixeira, L. D.; Freitas, S. M.; Otero-Gonzalez, A. J.; Gonçalves, S.; Santos, N. C.; Franco, O. L. Structural and Functional Evaluation of the Palindromic Alanine-Rich Antimicrobial Peptide Pa-MAP. *Biochim. Biophys. Acta* **2016**, *1858* (7 Pt A),1488–1498.

Moore, K. S.; Wehrli, S.; Roder, H.; Rogers, M.; Forrest, J. N.; McCrimmon, D.; Zasloff, M. Squalamine: An Aminosterol Antibiotic from the Shark. *Proc. Natl. Acad. Sci. U.S.A.* **1993**, *90*, 1354–1358.

Moskowitz, R. W. Role of Collagen Hydrolysate in Bone and Joint Disease. *Semin. Arthritis Rheum.* **2000**, *30*, 87–99.

Mozaffarian, D.; Rimm, E. B. Fish Intake, Contaminants, and Human Health: Evaluating the Risks and the Benefits. *JAMA* **2006**, *296*, 1885–1899.

Muncaster, S.; Kraakman, K.; Gibbons, O.; Mensink, K.; Forlenza, M.; Jacobson, G.; Bird, S. Antimicrobial Peptides within the Yellowtail Kingfish (*Seriola lalandi*). *Dev. Comp. Immunol.* **2018**, *80*, 67–80.

Murthy, L. N.; Rao, B. M.; Asha, K.; Prasad, M. M. Extraction and Quality Evaluation of Yellowfin Tuna Bone Powder. *Fish. Technol.* **2014**, *51*, 38–42.

Muthumari, K.; Anand, M.; Maruthupandy, M. Collagen Extract from Marine Finfish Scales as a Potential Mosquito Larvicide. *Protein J.* **2016**, *35*, 391–400.

Muyonga, J. H.; Cole, C. G. B.; Duodu, K. G. Characterisation of Acid Soluble Collagen from Skins of Young and Adult Nile Perch (*Lates niloticus*). *Food Chem.* **2004**, *85*, 81–89.

Nagai, T.; Suzuki, N. Isolation of Collagen from Fish Waste Material—Skin, Bone, and Fins. *Food Chem.* **2000**, *68*, 277–281.

Nagai, T.; Izumi, M.; Ishii, M. Fish Scale Collagen. Preparation and Partial Characterization. *Int. J. Food Sci. Technol.* **2004**, *39*, 239–244.

Najafian, L.; Babji, A. S. A Review of Fish-Derived Antioxidant and Antimicrobial Peptides: Their Production, Assessment, and Applications. *Peptides* **2012**, *33*, 178–185.

Nalinanon, S.; Benjakul, S.; Visessanguan, W.; Kishimura, H. Tuna Pepsin: Characteristics and Its Use for Collagen Extraction from the Skin of Deep-Sea Redfish (*Sebastes mentella*). *J. Food Sci.* **2007**, *73*, C413–C419.

Naqash, S. Y.; Nazeer, R. A. Antioxidant Activity of Hydrolysates and Peptide Fractions of *Nemipterus japonicus* and *Exocoetus volitans* Muscle. *J. Aquat. Food Prod. Technol.* **2010**, *19*, 180–192.

Naqash, S. Y.; Nazeer, R. A. Evaluation of Bioactive Properties of Peptide Isolated from *Exocoetus volitans* Backbone. *Int. J. Food Sci. Technol.* **2011**, *46*, 37–43.

Naqash, S. Y.; Nazeer, R. A. In Vitro Antioxidant and Antiproliferative Activities of Bioactive Peptide Isolated from *Nemipterus Japonicus* Backbone. *Int. J. Food Prop.* **2012**, *15*, 1200–1211.

Nasri, R.; Amor, I. B.; Bougatef, A.; Nedjar-Arroume, N.; Dhulster, P.; Gargouri, J.; Châabouni, M. K.; Nasri, M. Anticoagulant Activities of Goby Muscle Protein Hydrolysates. *Food Chem.* **2012a**, *133*, 835–841.

Nasri, R.; Bougatef, A.; Khaled, H. B.; Nedjar-Arroume, N.; Chaâbouni, M. K.; Dhulster, P.; Nasri, M. Antioxidant and Free Radical-Scavenging Activities of Goby (*Zosterisessor ophiocephalus*) Muscle Protein Hydrolysates Obtained by Enzymatic Treatment. *Food Biotechnol.* **2012b**, *26*, 266–279.

Nasri, R.; Jridi, M.; Lassoued, I.; Jemil, I.; Salem, R. B.; Nasri, M.; Karra-Châabouni, M. The Influence of the Extent of Enzymatic Hydrolysis on Antioxidative Properties and ACE-Inhibitory Activities of Protein Hydrolysates from Goby (*Zosterisessor ophiocephalus*) Muscle. *Appl. Biochem. Biotechnol.* **2014**, *173*, 1121–1134.

Nasri, R.; Abdelhedi, O.; Jemil, I.; Daoued, I.; Hamden, K.; Kallel, C.; Elfeki, A; Lamri-Senhadji, M.; Boualga, A.; Nasri, M.; Karra-Châabouni, M. Ameliorating Effects of Goby Fish Protein Hydrolysates on High-Fat-High-Fructose Diet-Induced Hyperglycaemia, Oxidative Stress and Deterioration of Kidney Function in Rats. *Chem. Biol. Interact.* **2015**, *242*, 71–80.

Nasri, R.; Abdelhedi, O.; Jemil, I.; Amor, I. B.; Elfeki, A.; Gargouri, J.; Boualga, A.; Karra Chaabouni, M.; Nasria, M. Preventive Effect of Goby Fish Protein Hydrolysates on Hyperlipidemia and Cardiovascular Disease in Wistar Rats Fed a High-Fat/Fructose Diet. *RSC Adv.* **2018**, *8*, 9383–9393.

Navarro-Garcia, G.; Pacheco-Aguilar, R.; Bringas-Alvarado, L.; Ortega-Garcia, J. Characterization of the Lipid Composition and Natural Antioxidants in the Liver Oil of *Dasyatis brevis* and *Gymnura marmorata* Rays. *Food Chem.* **2004**, *87*, 89–96.

Nazeer, R. A.; Deeptha, R. Antioxidant Activity and Amino Acid Profiling of Protein Hydrolysates from the Skin of *Sphyraena barracuda* and *Lepturacanthus savala*. *Int. J. Food Prop.* **2013**, *16*, 500–511.

Nazeer, R. A.; Kulandai, K. A. Evaluation of Antioxidant Activity of Muscle and Skin Protein Hydrolysates from Giant Kingfish, *Caranx ignobilis* (Forsskål, 1775). *Food Sci. Technol.* **2012**, *47*, 274–281.

Nazeer, R. A.; Kumar, N. S. S.; Naqash, S. Y.; Radhika, R. Lipid Profiles of Threadfin Bream (*Nemipterus japonicus*) Organs. *Indian J. Geo-Mar. Sci.* **2009**, *38*, 461–463.

Nazeer, R.; Deeptha, R.; Jaiganesh, R.; Sampathkumar, N. S.; Shabeena, N. Radical Scavenging Activity of Seela (*Sphyraena barracuda*) and Ribbon Fish (*Lepturacanthus savala*) Backbone Protein Hydrolysates. *Int. J. Pept. Res. Ther.* **2011,** *17,* 209–216.

Nemati, M.; Huda, N.; Ariffin, F. Development of Calcium Supplement from Fish Bone Wastes of Yellowfin Tuna (*Thunnus albacares*) and Characterization of Nutritional Quality. *Int. Food Res. J.* **2017,** *24,* 2419–2426.

Ngo, D. H.; Qian, Z. J.; Ryu, B.; Park, J. W.; Kim, S. K. In Vitro Antioxidant Activity of a Peptide Isolated from Nile Tilapia (*Oreochromis niloticus*) Scale Gelatin in Free Radical-Mediated Oxidative Systems. *J. Funct. Foods* **2010,** *2,* 107–117.

Ngo, D. H.; Ryu, B.; Vo, T. S.; Himaya, S. W. A.; Wijesekara, I.; Kim, S. K. Free Radical Scavenging and Angiotensin-I Converting Enzyme Inhibitory Peptides from Pacific Cod (*Gadus macrocephalus*) Skin Gelatin. *Int. J. Biol. Macromol.* **2011,** *49,* 1110–1116.

Ngo, D.; Vo, T.; Ngo, D.; Wijesekara, I.; Kima, S. Biological Activities and Potential Health Benefits of Bioactive Peptides Derived from Marine Organisms. *Int. J. Biol. Macromol.* **2012,** *51,* 378–383.

Ngo, D.; Kang, K.; Jung, W.; Byun, H.; Kim, S. Protective Effects of Peptides from Skate (*Okamejei kenojei*) Skin Gelatin against Endothelial Dysfunction. *J. Funct. Foods* **2014,** *10,* 243–251.

Nikoo, M.; Benjakul, S.; Xu, X. Antioxidant and Cryoprotective Effects of *Amur sturgeon* Skin Gelatin Hydrolysate in Unwashed Fish Mince. *Food Chem.* **2015,** *181,* 295–303.

Noitup, P.; Garnjanagoonchorn, W.; Morrissey, M. T. Fish Skin Type I Collagen: Characteristic Comparison of Albacore Tuna (*Thunnus alalunga*) and Silver-Line Grunt (*Pomadasys kaakan*). *J. Aquat. Food Prod. Technol.* **2005,** *14,* 17–28.

Nomura, A.; Noda, N.; Maruyama, S. Purification of Angiotensin I-Converting Enzyme Inhibitors in Pelagic Thresher *Alopias pelagicus* Muscle Hydrolysate and Viscera Extracts. *Fish. Sci.* **2002,** *68,* 954–956.

Normah, I.; Nur-Hani Suryati, M. Z. Isolation of Threadfin Bream (*Nemipterus japonicus*) Waste Collagen Using Natural Acid from Calamansi (*Citrofortunella microcarpa*) Juice. *Int. Food Res. J.* **2015,** *22,* 2294–2301.

Nurdiani, R.; Dissanayake, M.; Street, W. E.; Donkor, O. N.; Singh, T. K.; Vasiljevic, T. Sustainable Use of Marine Resources—Turning Waste into Food Ingredients. *Int. J. Food Sci. Technol.* **2015,** *50,* 2329–2339.

Nurdiani, R.; Dissanayake, M.; Street, W. E.; Donkor, O. N.; Singh, T. K.; Vasiljevic, T. In Vitro Study of Selected Physiological and Physicochemical Properties of Fish Protein Hydrolysates from 4 Australian Fish Species. *Int. Food Res. J.* **2016,** *23,* 2029–2040.

Nurdiani, R.; Vasiljevic, T.; Yeager, T.; Singh, T. K.; Donkor, O. N. Bioactive Peptides with Radical Scavenging and Cancer Cell Cytotoxic Activities Derived from Flathead (*Platycephalus fuscus*) By-products. *Eur. Food Res. Technol.* **2017,** *243,* 627–637.

Ogawa, M.; Moody, M. W.; Portier, R. J.; Bell, J.; Schexnayder, M.; Losso, J. N. Biochemical Properties of Black Drum and Sheepshead Sea Bream Skin Collagen. *J. Agric. Food Chem.* **2003,** *51,* 8088–8092.

Olgunoglu, I. A. Review on Omega-3 (n-3) Fatty Acids in Fish and Seafood. *J. Biol. Agric. Healthc.* **2017,** *7,* 37–45.

Oliveira, V. M.; Assis, C. R. D.; Herculano, P. N.; Cavalcanti, M. T. H.; Bezerra, R. S.; Porto, B. A. L. F. Collagenase from Smooth Weakfish: Extraction, Partial Purification,

Characterization and Collagen Specificity Test for Industrial Application. *Inst. Pesca, São Paulo* **2017**, *43*, 52–64.

Oren, Z.; Shai, Y. A Class of Highly Potent Antibacterial Peptides Derived from Pardaxin, a Pore-Forming Peptide Isolated from Moses Sole Fish *Pardachirus marmoratus*. *Eur. J. Biochem.* **1996**, *237*, 303–310.

Oscarsson, J.; Hurt-Camejo, E. Omega-3 Fatty Acids Eicosapentaenoic Acid and Docosahexaenoic Acid and Their Mechanisms of Action on Apolipoprotein B-Containing Lipoproteins in Humans: A Review. *Lipids Health Dis.* **2017**, *16*, 149.

Osterud, B.; Elvevoll, E.; Barstad, H.; Brox, J.; Halvorsen, H.; Lia, K.; Olsen, J. O.; Olsen, R. L.; Sissener, C.; Rekdal, O; Vognild, E. Effect of Marine Oils Supplementation on Coagulation and Cellular Activation in Whole Blood. *Lipids* **1995**, *30*, 1111–1118.

Pachaiyappan, A.; Sadhasivam, G.; Arumugam, S.; Muthuvel, A. Bioprospecting the Enzymatic and Anticancer Potential of Spine Secretions of Marine Catfish (*Plotosus lineatus*). *JBAPN* **2015**, *5*, 406–418.

Pampanina, D. M.; Larssena, E.; Provana, F.; Sivertsvik, M.; Ruoff, Q. P.; Sydnes, M. O. Detection of Small Bioactive Peptides from Atlantic Herring (*Clupea harengus* L.). *Peptides* **2012**, *34*, 423–426.

Pamungkas, B. F.; Supriyadi; Murdiati, A.; Indrati, R. Characterization of the Acid- and Pepsin-Soluble Collagens from Haruan (*Channa striatus*) Scales. *Pak. J. Nutr.* **2019**, *18*, 324–332.

Pan, X.; Zhao, Y.; Hu, F.; Wang, B. Preparation and Identification of Antioxidant Peptides from Protein Hydrolysate of Skate (*Raja porosa*) Cartilage. *J. Funct. Foods* **2016a**, *25*, 220–230.

Pan, X.; Zhao, Y.; Hu, F.; Chi, C.; Wang, B. Anticancer Activity of a Hexapeptide from Skate (*Raja porosa*) Cartilage Protein Hydrolysate in HeLa Cells. *Mar. Drugs* **2016b**, *14*, 153.

Pangestuti, R.; Kim, S. Bioactive Peptide of Marine Origin for the Prevention and Treatment of Non-communicable Diseases. *Mar. Drugs* **2017**, *15*, 67.

Park, C. B.; Lee, J. H.; Park, I. Y.; Kim, M. S.; Kim, S. C. A Novel Antimicrobial Peptide from the Loach, *Misgurnus anguillicaudatus*. *FEBS Lett.* **1997**, *411*, 173–178.

Patra, D.; Sandell, L. J. Antiangiogenic and Anticancer Molecules in Cartilage. *Expert Rev. Mol. Med.* **2012**, *14*, e10.

Pellizzon, M.; Buison, A.; Ordiz, F. Jr.; Ana, L. S.; Jen, K. L. C. Effects of Dietary Fatty Acids and Exercise on Body-Weight Regulation and Metabolism in Rats. *Obes. Res.* **2002**, *10*, 947–955.

Peyronel, D.; Artaud, J.; Iatrides, M, C.; Chevalier, J. L. Fatty Acid and Squalene Compositions of Mediterranean Centrophorus SPP Egg and Liver Oils in Relation to Age. *Lipids* **1984**, *19*, 643–648.

Phantura, P.; Benjakul, S.; Visessanguan, W.; Roytrakul, S. Use of Pyloric Caeca Extract for the Production of Gelatin Hydrolysate with Antioxidative Activity. *LWT—Food Sci. Technol.* **2010**, *43*, 86–97.

Picot, L.; Bordenave, S.; Didelot, S.; Fruitier-Arnaudin, I.; Sannier, F.; Thorkelsson, G.; Bergé, J. P.; Guérard, F.; Chabeaud, A.; Piot, J. M. Antiproliferative Activity of Fish Protein Hydrolysates on Human Breast Cancer Cell Lines. *Process Biochem.* **2006**, *41*, 1217–1222.

Prithiviraj, N.; Sasikala, R.; Annadurai, D. Bioactive Properties of the Stone Fish *Synanceia horrida* (Thomas, 1984) Spine Venom. *Int. J Pharm. Biol. Arch.* **2012**, *3*, 1217–1221.

Prithiviraj, N.; Annadurai, D.; Kumaresan, S. Biological Properties of the Orange Banded Sting Fish Spine Venom *Choridactylus multibarbus* (Richardson, 1848) from Parangipettai Coast, Southeast Coast of India. *Int. J. Curr. Res. Chem. Pharm. Sci.* **2015**, *2*, 35–45.

Priya, K. M.; Khora, S. S. Antimicrobial, Hemolytic and Cytotoxic activities of the Puffer Fish *Arothron hispidus* from the Southeast Coast of India. *Int. J. Drug Dev. Res.* **2013**, *5*, 317–322.

Priya, E. R.; Ravichandran, S.; Ghosh, S. Pharmacological Evaluation of Bioactive Peptide from *Lagocephalus spadiceus*. *World J. Pharm. Pharm. Sci.* **2016**, *5*, 1004–1015.

Qian, Z. J.; Je, J. Y.; Kim, S. K. Antihypertensive Effect of Angiotensin I Converting Enzyme-Inhibitory Peptide from Hydrolysates of Bigeye Tuna Dark Muscle, *Thunnus obesus*. *J. Agric. Food Chem.* **2007**, *55*, 8398–8403.

Rabiei, S.; Nikoo, M.; Rezaei, M.; RaFleian-Kopaei, M. Marine-Derived Bioactive Peptides with Pharmacological Activities—A Review. *J. Clin. Diagn. Res.* **2017**, *11*, KE01–KE06.

Raj, M. M.; Bragadeeswaran, S.; Suguna, A. Studies on Haemolytic Properties of Puffer Fishes from South East Coast of India. *Int. Lett. Nat. Sci.* **2014**, *30*, 11–18.

Raj, M. M.; Bragadeeswaran, S.; Suguna, A.; Siva, M. U. Studies on Antimicrobial Activity and Brine Shrimp Lethality of Crude Samples of Six Different Species of Puffer Fishes. *J. Coast Life Med.* **2015**, *3*, 515–517.

Rajani, N.; Alka, M. To study the Ethano-medicinal Importance of Food Fish Used by Localite of Drug. *IOSR J. Environ. Sci. Toxicol. Food Technol.* **2015**, *1*, 38–40.

Rajapakse, N.; Jung, W.; Mendis, E Moon, S.; Kim, S. A Novel Anticoagulant Purified from Fish Protein Hydrolysate Inhibits Factor XIIa and Platelet Aggregation. *Life Sci.* **2005**, *76*, 2607–2619.

Ranathunga, S.; Rajapakse, N.; Kim, S. Purification and Characterization of Antioxidative Peptide Derived from Muscle of Conger Eel (*Conger myriaster*). *Eur. Food Res. Technol.* **2006**, *222*, 310–315.

Rao, G. N. Physico-chemical, Functional and Antioxidant Properties of Roe Protein Concentrates from *Cyprinus carpio* and *Epinephelus tauvina*. *J. Food Pharm. Sci.* **2014**, *2*, 15–22.

Rao, M. K. G.; Acaya, K. T. Antioxidant Activity of Squalene. *J. Am. Oil Chem. Soc.* **1968**, *45*, 296.

Ratnasari, I.; Yuwono, S. S.; Nusyam, H.; Widjanarko, S. B. Extraction and Characterization of Gelatin from Different Fresh Water Fishes as Alternative Sources of Gelatin. *Int. Food Res. J.* **2013**, *20*, 3085–3091.

Remme, J. F.; Larsen, W. E.; Stoknes, I. S. Bioactive Lipids in Deep-Sea Sharks. Report Å0510. More Research, Alesund, Report Å0510. **2005**. file:///C:/Users/santhanam/Downloads/%C3%85O510%20(1).pdf.

Rethinam, S; Nivedita, P; Hemalatha, T; Vedakumari, S, W.; Sastry, T. P. A Possible Wound Dressing Material from Marine Food Waste. *Int. J. Artif. Organs* **2016**, *39*, 509–517.

Ritchie, K. B; Schwarz, M; Mueller, J; Lapacek, V. A.; Merselis, D.; Walsh, C. J.; Luer, C. A. Survey of Antibiotic-Producing Bacteria Associated with the Epidermal Mucus Layers of Rays and Skates. *Front. Microbiol.* **2017**. https://doi.org/10.3389/fmicb.2017.01050.

Rubio-Rodríguez, N.; De Diego, S. M.; Beltrán, S.; Jaime, I.; Sanz, M. T.; Rovira, J. Supercritical Fluid Extraction of Fish Oil from Fish By-products: A Comparison with Other Extraction Methods. *J. Food Eng.* **2012**, *109*, 238–248.

Sadowska, M.; Kołodziejska, I.; Niecikowska, C. Isolation of Collagen from the Skins of Baltic Cod (*Gadus morhua*). *Food Chem.* **2003**, *81*, 257–262.

Sae-Leaw, T.; Karnjanapratum, S.; O'Callaghan, Y. C.; O'Keeffe, M. B.; FitzGerald, R. J.; O'Brien, N. M.; Benjakul, S. Purification and Identification of Antioxidant Peptides from Gelatin Hydrolysate of Seabass Skin. *J. Food Biochem.* **2017**, *41*, e12350.

Saitoh, T.; Seto, Y.; Fujikawa, Y.; Iijima, N. Distribution of Three Isoforms of Antimicrobial Peptide, Chrysophsin-1, -2 and -3, in the Red Sea Bream, *Pagrus (Chrysophrys) major*. *Anal. Biochem.* **2019**, *566*, 13–15.

Salaudeen, M. M. Quality Analysis of Dried Cod (*Gadus morhua*) Heads along the Value Chain from Iceland to Nigeria. United Nations University Fisheries Training Programme, Iceland [Final Project]. **2014**. http://www.unuftp.is/static/fellows/document/mutiat13prf.pdf.

Samidurai, K.; Mathew, N. Mosquito Larvicidal and Ovicidal Activity of Puffer Fish Extracts against *Anopheles stephensi, Culex quinquefasciatus* and *Aedes aegypti* (Diptera: Culicidae). *Trop. Biomed.* **2013**, *30*, 27–35.

Santhanam, R. Fishery Byproduct. In: *Fisheries Science*; Daya Publishing House, Delhi, **1990**; pp 145–147.

Sathivel, S.; Bechtel, P. J.; Babbitt, J.; Smiley, S.; Crapo, C.; Reppond, K. D.; Prinyawiwatkul, W. Biochemical and Functional Properties of Herring (*Clupea harengus*) Byproduct Hydrolysates. *J. Food Sci.* **2013**, *68*, 2196–2200.

Sayari, N.; Sila A Haddar, A.; Balti, R.; Ellouz-Chaabouni, S.; Bougatef, A. Valorisation of Smooth Hound (*Mustelus mustelus*) Waste Biomass through Recovery of Functional, Antioxidative and Antihypertensive Bioactive Peptides. *Environ. Sci. Pollut. Res. Int.* **2016**, *23*, 366–376.

Sego, S., Cod Liver Oil: An Antibacterial, Antifungal Agent. 2017. https://www.clinicaladvisor.com/home/features/alternative-meds-update/cod-liver-oil-an-antibacterial-antifungal-agent/.

Senaratne, L. S.; Park, P.; Kim, S. Isolation and Characterization of Collagen from Brown Backed Toadfish (*Lagocephalus gloveri*) Skin. *Bioresour. Technol.* **2006**, *97*, 191–197.

Seo, J. K.; Lee, M. J.; Go, H. J.; Park, T. H.; Park, N. G. Purification and Characterization of YFGAP, a GAPDH-Related Novel Antimicrobial Peptide, from the Skin of Yellowfin Tuna, *Thunnus albacares*. *Fish Shellfish Immunol.* **2012**, *33*, 743–752.

Seo, J. K.; Lee, M. J.; Go, H. J.; Kim, Y. J.; Park, N. G. Antimicrobial Function of the GAPDH Related Antimicrobial Peptide in the Skin of Skipjack Tuna, *Katsuwonus pelamis. Fish Shellfish Immunol.* **2014**, *36*, 571–581.

Shafri, M. A. M.; Manan, M. J. A. Therapeutic Potential of the Haruan (*Channa striatus*). From Food to Medicinal Uses. *Mal. J. Nutr.* **2012**, *18*, 125–136.

Shahidi, F. Nutraceuticals and Bioactives from Seafood Byproducts. In *Advances in Seafood Byproducts. Conference Proceedings, Proceedings of the 2nd International Seafood Byproduct Conference*; Bechtel, P. J. Ed.; Anchorage, Alaska, USA, **2002**; pp 247–264.

Shahidi, F. *Maximising the Value of Marine By-products: Technology and Engineering*. Woodhead Publishing: Cambridge, United Kingdom, 2006; p 560.

Shahidi, F.; Abuzaytoun, R. Chitin, Chitosan, and Co-products: Chemistry, Production, Applications, and Health Effects. *Adv. Food Nutr. Res.* **2005**, *49*, 93–135.

Shahidi, F.; Varatharajan, V.; Peng, H.; Senadheera, R. Utilization of Marine By-products for the Recovery of Value-Added Products. *J. Food Bioact.* **2019**, *6*, 10–61.

Sheriff, S. A.; Balasubramanian, S.; Baranitharan, R.; Ponmurugan, P. Synthesis and In Vitro Antioxidant Functions of Protein Hydrolysate from Backbones of *Rastrelliger kanagurta* by Proteolytic Enzymes. *Saudi J. Biol Sci.* **2014**, *21*, 19–26.

Shi, P.; Liu, M.; Fan, F.; Yu, C.; Lu, W.; Du, M. Characterization of Natural Hydroxyapatite Originated from Fish Bone and Its Biocompatibility with Osteoblasts. *Mater. Sci. Eng. C: Mater. Biol. Appl.* **2018**, *90*, 706–712.

Silva, T. H.; Moreira-Silva, J.; Marques, A. L. P.; Domingues, A.; Bayon, Y.; Reis, R. L. Marine Origin Collagens and Its Potential Applications. *Mar. Drugs* **2014**, *12*, 5881–5901.

Simpson, B. K. *Food Biochemistry and Food Processing: Technology & Engineering;* John Wiley & Sons: Hoboken, NJ, **2012**; p 912.

Singh, P.; Benjakul, S.; Maqsood, S.; Kishimura, H. Isolation and Characterisation of Collagen Extracted from the Skin of Striped Catfish (*Pangasianodon hypophthalmus*). *Food Chem.* **2011**, *124*, 97–105.

Sionkowska, A.; Kozłowska, J.; Skorupska, M.; Michalska, M. Isolation and Characterization of Collagen from the Skin of *Brama australis*. *Int. J. Biol. Macromol.* **2015**, *80*, 605–609.

Song, R.; Wei, R.; Zhang, B.; Yang, Z.; Wang, S. Antioxidant and Antiproliferative Activities of Heated Sterilized Pepsin Hydrolysate Derived from Half-Fin Anchovy (*Setipinna taty*). *Mar. Drugs* **2011**, *9*, 1142–1156.

Song, R.; Wei, R.; Luo, H.; Wang, D. Isolation and Characterization of an Antibacterial Peptide Fraction from the Pepsin Hydrolysate of Half-Fin Anchovy (*Setipinna taty*). *Molecules* **2012**, *17*, 2980–2991.

Song, H.; Meng, M.; Cheng, X; Li, B.; Wang, C. The Effect of Collagen Hydrolysates from Silver Carp (*Hypophthalmichthys molitrix*) Skin on UV-Induced Photoaging in Mice: Molecular Weight Affects Skin Repair. *Food Funct.* **2017**, *8*, 1538–1546.

Sotelo, C. G.; Comesaña, M. B.; Ariza, P. R.; Pérez-Martín, R. I. Characterization of Collagen from Different Discarded Fish Species of the West Coast of the Iberian Peninsula. *J. Aquat. Food Prod. Technol.* **2016**, *25*, 388–399.

Souissi, N.; Bougatef, A.; Triki-Ellouz, Y.; Nasri, M. Biochemical and Functional Properties of Sardinella (*Sardinella aurita*) By-product Hydrolysates. *Food Technol. Biotechnol.* **2007**, *45*, 187–194.

Squires, E. J.; Haard, N. F.; Feltham, L. A. W. Pepsin Isozymes from Greenland Cod, *Gadus ogac*. 1. Purification and Physical Properties. *Can. J. Biochem. Cell Biol.* **1986**, *65*, 205–209.

Su, Y. Isolation and Identification of Pelteobagrin, a Novel Antimicrobial Peptide from the Skin Mucus of Yellow Catfish (*Pelteobagrus fulvidraco*). *Comp. Biochem. Physiol. B: Biochem. Mol. Biol.* **2011**, *158*, 149–154.

Subramanian, S.; Ross, N. W.; MacKinnon, S. L. Myxinidin, a Novel Antimicrobial Peptide from the Epidermal Mucus of Hagfish, *Myxine glutinosa* L. *Mar. Biotechnol. N.Y.* **2009**, *11*, 748–757.

Sun, L.; Wu, S.; Zhou, L.; Wang, F.; Lan, X.; Sun, J.; Tong, Z.; Liao, D. Separation and Characterization of Angiotensin I Converting Enzyme (ACE) Inhibitory Peptides from *Saurida elongata* Proteins Hydrolysate by IMAC-Ni2. *Mar. Drugs* **2017**, *15*, 29.

Sun, J.; Zhang, J.; Zhao, D.; Changhu Xue, C.; Zhen Liu, Z.; Xiangzhao Mao, X. Characterization of Turbot (*Scophthalmus maximus*) Skin and the Extracted Acid-Soluble Collagen. *J. Ocean Univ. China* **2019**, *18*, 687–692.

Sung-Hoi, H. Feeding Habits of *Syngnathus schlegeli* in Eelgrass (*Zostera marina*) Bed in Kwangyang Bay. *Korean J. Fish. Aquat. Sci.* **1997**, *30*, 896–902.

Suntornsaratoon, P.; Charoenphandhu, N.; Krishnamra, N. Fortified Tuna Bone Powder Supplementation Increases Bone Mineral Density of Lactating Rats and their Offspring. *J. Sci. Food Agric.* **2018**, *98*, 2027–2034.

Taheri, B.; Mohammadi, M.; Nabipour, I.; Momenzadeh, N.; Roozbehani, M. Identification of Novel Antimicrobial Peptide from Asian Sea Bass (*Lates calcarifer*) by In Silico and Activity Characterization. *PLoS One* **2018**, *13*, e0206578.

Tanuja, S.; Viji, P.; Zynudheen, A. A.; Joshy, C. G. Composition, Functional Properties and Antioxidative Activity of Hydrolysates Prepared from the Frame Meat of Striped Catfish (*Pangasianodon hypophthalmus*). *Egypt. J. Biol.* **2012**, *14*, 28–36.

Tao, J.; Zhao, Y.; Chi, C.; Wang, B. Bioactive Peptides from Cartilage Protein Hydrolysate of Spotless Smoothhound and Their Antioxidant Activity In Vitro. *Mar. Drugs* **2018**, *16*, 100.

Teixeira, L. D.; Silva, O. N.; Migliolo, L.; Fensterseifer, I. C. M.; Franco, O. L. In Vivo Antimicrobial Evaluation of an Alanine-Rich Peptide Derived from *Pleuronectes americanus*. *Peptides* **2013**, *42*, 144–148.

Thiansilakul, Y.; Benjakul, S.; Shahidi, F. Antioxidative Activity of Protein Hydrolysate from Round Scad Muscle Using Alcalase and Flavourzyme. *J. Food Chem.* **2007**, *31*, 266–287.

Thuanthong, M.; De Gobba, C.; Sirinupong, N; Youravong, W.; Otte, J. Purification and Characterization of Angiotensin-Converting Enzyme-Inhibitory Peptides from Nile Tilapia (*Oreochromis niloticus*) Skin Gelatine Produced by an Enzymatic Membrane Reactor. *J. Funct. Foods* **2017**, *36*, 243–254.

Thuy, L. T. M.; Okazaki, E.; Osako, K. Isolation and Characterization of Acid-Soluble Collagen from the Scales of Marine Fishes from Japan and Vietnam. *Food Chem.* **2014**, *149*, 264–270.

Toppe, J.; Albrektsen, S.; Hope, B.; Aksnes, A. Chemical Composition, Mineral Content and Amino Acid and Lipid Profiles in Bones from Various Fish Species. *Comp. Biochem. Physiol. B: Biochem. Mol. Biol.* **2007**, *146*, 395–401.

Tvete, T.; Haugan, K. Purification and Characterization of a 630 kDa Bacterial Killing Metalloprotease (KilC) Isolated from Plaice *Pleuronectes platessa* (L.), Epidermal Mucus. *J. Fish Dis.* **2008**, *31*, 343–352.

Tylingo, R.; Mania, S.; Panek, A.; Piątek, R.; Pawłowicz, R. Isolation and Characterization of Acid Soluble Collagen from the Skin of African Catfish (*Clarias gariepinus*), Salmon (*Salmo salar*) and Baltic Cod (*Gadus morhua*). *J. Biotechnol. Biomater.* **2016**, *6*, 234.

Vannuccini, S. *Shark Utilization, Marketing, and Trade*; Food & Agriculture Org., Business & Economics **1999**; 470 p.

Vázquez, J. A.; Blanco, M.; Massa, A. E.; Amado, I. R.; Pérez-Martín, R. I. Production of Fish Protein Hydrolysates from *Scyliorhinus canicula* Discards with Antihypertensive and Antioxidant Activities by Enzymatic Hydrolysis and Mathematical Optimization Using Response Surface Methodology. *Mar. Drugs* **2017**, *15*, E306.

Veeruraj, A.; Arumugam, M.; Ajithkumar, T.; Balasubramanian, T. Isolation and Characterization of Drug Delivering Potential of Type-I Collagen from Eel Fish, *Evenchelys macrura*. *J. Mater. Sci. Mater. Med.* **2012**, *23*, 1729–1738.

Venkatesan, J.; Sukumaran Anil, S.; Kim, S.; Shim, M. S. Marine Fish Proteins and Peptides for Cosmeceuticals: A Review. *Mar. Drugs* **2017**, *15*, 143.

Vennila, R.; Kumar, K. R.; Kanchana, S.; Arumugam, M.; Vijayalakshmi, S.; Balasubramaniam, T. Preliminary Investigation on Antimicrobial and Proteolytic Property of the Epidermal Mucus Secretion of Marine Stingrays. *Asian Pac. J. Trop. Biomed.* **2011**, *1*, 239–243.

Venugopal, V.; Kumaran, A. K.; Chatterjee, N. S.; Kumar, S.; Kavilakath, S.; Nair, J. R.; Mathew, S. Biochemical Characterization of Liver Oil of *Echinorhinus brucus* (Bramble Shark) and Its Cytotoxic Evaluation on Neuroblastoma Cell Lines (SHSY-5Y). *Scientifica* **2016**, 6 p; ID 6294030.

Viet, B. T. N. T.; Ohshima, T. The Stability of Bioactive Compounds in Yellowstripe Scad (*Selaroides leptolepis*) under Subatmospheric Pressure Storage. *Int. J. Res Agric. Food Sci.* **2014**, *2*, 23–28.

Vishnu, K. V.; Kumar, K. K. A.; Asha, K. K.; Remyakumari, K. R.; Ganesan, B.; Anandan, R.; Sekhar Chatterjee, N. S.; Mathew, S. Protective Effects of *Echinorhinus brucus* Liver Oil against Induced Inflammation and Ulceration in Rats. *Fish. Technol.* **2015**, *52*, 1–6.

Vishnu, K. V.; Ajeeshkumar, K. K.; Remyakumari, K. R.; Ganesan, B.; Niladri, S. C.; Lekshmi, R. G. K.; Shyni, K.; Mathew, S. Gastroprotective Effect of Sardine Oil (*Sardinella longiceps*) against HCl/Ethanol Induced Ulceration in Wistar Rats. *Int. J. Fish. Aquat. Stud.* **2017**, *5*, 118–124.

Vo, T.; Ngo, J. D.; Kim, J.; Ryu, B.; Kim, S. An Antihypertensive Peptide from Tilapia Gelatin Diminishes Free Radical Formation in Murine Microglial Cells. *J. Agric. Food Chem.* **2011**, *59* (22), 12193–12197.

Walsh, C. J.; Luer, C. A.; Bodine, A. B.; Smith, C. A.; Cox, H. L.; Noyes, D. R.; Maura, G. Elasmobranch Immune Cells as a Source of Novel Tumor Cell Inhibitors: Implications for Public Health. *Integr. Comp. Biol.* **2006**, *46*, 1072–1081.

Walsh, C. J.; Luer, C. A.; Yordy, J. E.; Cantu, T.; Miedema, J.; Leggett, S. R.; Leigh, B.; Adams, P.; Ciesla, M.; Bennett, C.; Bodine, A. B. Epigonal Conditioned Media from Bonnethead Shark, *Sphyrna tiburo*, Induces Apoptosis in a T-Cell Leukemia Cell Line, Jurkat E6-1. *Mar. Drugs* **2013**, *11*, 3224–3257.

Wang, W.; Tao, R.; Tong, Z.; Ding, Y.; Kuang, R.; Zhai, S.; Liu, J.; Ni, L. Effect of a Novel Antimicrobial Peptide Chrysophsin-1 on Oral Pathogens and *Streptococcus mutans* Biofilms. *Peptides* **2012**, *33*, 212–219.

Wang, B.; Wang, Y.; Chi, C.; Luo, H.; Deng, S.; Ma, J. Isolation and Characterization of Collagen and Antioxidant Collagen Peptides from Scales of Croceine Croaker (*Pseudosciaena crocea*). *Mar. Drugs* **2013**, *11*, 4641–4661.

Wang, T. Y.; Hsieh, C. H.; Hung, C. C.; Jao, C. L.; Chen, M. C.; Hsu, K. C. Fish Skin Gelatin Hydrolysates as Dipeptidyl Peptidase IV Inhibitors and Glucagon-Like Peptide 1 Stimulators Improve Glycaemic Control in Diabetic Rats: A Comparison between Warm- and Cold-Water Fish. *J. Funct.* **2015**, *19*, 330–340.

Wang, X.; Yu, H.; Xing, R.; Li, P. Characterization, Preparation, and Purification of Marine Bioactive Peptides. *BioMed. Res. Int.* **2017**, ID 9746720, 16 p.

Wang, X.; Yu, H.; Xing, R.; Chen, X.; Liu, S.; Li, P. Optimization of Antioxidative Peptides from Mackerel (*Pneumatophorus japonicus*) Viscera. *Peer J.* **2018**, *6*, e4373.

Wang, X.; Yu, H.; Xing, R.; Liu, S.; Chen, X.; Li, P. Preparation and Identification of Antioxidative Peptides from Pacific Herring (*Clupea pallasii*) Protein. *Molecules* **2019a**, *24*, 17 p.

Wang, J.; Wei, R.; Song, R. Novel Antibacterial Peptides Isolated from the Maillard Reaction Products of Half-Fin Anchovy (*Setipinna taty*) Hydrolysates/Glucose and Their Mode of Action *in Escherichia coli. Mar. Drugs* **2019b**, *17*, 47.

Wei, P.; Zheng, H.; Shi, Z.; Li, D.; Xiang, Y. Isolation and Characterization of Acid-Soluble Collagen and Pepsin-Soluble Collagen from the Skin of Hybrid Sturgeon. *J. Wuhan Univ. Technol.—Mater. Sci.* **2019**, *34*, 950–959.

Whitehurst, R. J.; Oort, M. V. *Enzymes in Food Technology: Technology & Engineering*; John Wiley & Sons: Hoboken, NJ, **2009**, 384 p.

Wijesekara, I.; Qian, Z.; Ryu, B. M.; Ngo, D.; Kim, S. Purification and Identification of Antihypertensive Peptides from Seaweed Pipefish (*Syngnathus schlegeli*) Muscle Protein Hydrolysate. *Food Res. Int.* **2011**, *44*, 703–707.

Wu, H.; Chen, H.; Shiau, C. Free Amino Acid and Peptide as Related to Antioxidant Properties in Protein Hydrolysates of Mackerel (*Scomber austriasicus*). *Food Res. Int.* **2003**, *36*, 949–957.

Wu, Q. Q.; Li, T.; Wang, B.; Ding, G. F. Preparation and Characterization of Acid and Pepsin-Soluble Collagens from Scales of Croceine and Redlip Croakers. *Food Sci. Biotechnol.* **2015a**, *24*, 2003–2010.

Wu, S.; Feng, X.; Lan, X.; Xu, Y. Purification and Identification of Angiotensin-I Converting Enzyme (ACE) Inhibitory Peptide from Lizard Fish (*Saurida elongata*) Hydrolysate. *J. Funct. Foods* **2015b**, *13*, 295–299.

Wu, G.; Wang, J.; Luo, P.; Li, A.; Tian, S.; Jiang, H.; Zheng, Y.; Zhu, F.; Lu, Y.; Xia, Z. Hydrostatin-SN1, a Sea Snake-Derived Bioactive Peptide, Reduces Inflammation in a Mouse Model of Acute Lung Injury. *Front. Pharmacol.* **2017**. https://doi.org/10.3389/fphar.2017.00246.

Xie, J.; Ye, H. Y.; Luo, X. F. An Efficient Preparation of Chondroitin Sulfate and Collagen Peptides from Shark Cartilage. *Int. Food Res. J.* **2014**, *21*, 1171–1175.

Xu, L.; Dong, W.; Zhao, J.; Xu, Y. Effect of Marine Collagen Peptides on Physiological and Neurobehavioral Development of Male Rats with Perinatal Asphyxia. *Mar. Drugs* **2015**, *13*, 3653–3671.

Xu, S.; Yang, H.; Shen, L.; Li, G. Purity and Yield of Collagen Extracted from Southern Catfish (*Silurus meridionalis* Chen) Skin through Improved Pretreatment Methods. *Int. J. Food Prop.* **2017**, *20*, S141–S153.

Yaghoubzadeh, Z.; Ghadikolaii, F. P.; Kaboosi, H.; Safari, R.; Fattahi, E. Antioxidant Activity and Anticancer Effect of Bioactive Peptides from Rainbow Trout (*Oncorhynchus mykiss*) Skin Hydrolysate. *Int. J. Pept. Res. Ther.* **2019**. https://doi.org/10.1007/s10989-019-09869-5.

Yang, J. I.; Ho, H. Y.; Chu, Y. J.; Chow, C. J. Characteristic and Antioxidant Activity of Retorted Gelatin Hydrolysates from Cobia (*Rachycentron canadum*) Skin. *Food Chem.* **2008**, *110*, 128–136.

Yang, P.; Ke, H.; Hong, P.; Zeng, S.; Cao, W. Antioxidant Activity of Bigeye Tuna (*Thunnus obesus*) Head Protein Hydrolysate Prepared with Alcalase. *Int. J. Food Sci. Technol.* **2011**, *46*, 2460–2466.

Yang, P.; Jiang, Y.; Hong, P.; Cao, W. Angiotensin I Converting Enzyme Inhibitory Activity and Antihypertensive Effect in Spontaneously Hypertensive Rats of Cobia (*Rachycentron canadum*) Head Papain Hydrolysate. *Food Sci. Technol. Int.* **2013**, *19*, 209–215.

Yang, J.; Tang, J.; Liu, Y.; Wang, H.; Lee, S.; Yen, C.; Chang, H. Roe Protein Hydrolysates of Giant Grouper (*Epinephelus lanceolatus*) Inhibit Cell Proliferation of Oral Cancer Cells Involving Apoptosis and Oxidative Stress. *Biomed. Res. Int.* **2016**, *2016*, ID 8305073, 12 p.

Yu, D.; Chi, C.; Wang, B.; Dinng, G.; Li, Z. Characterization of Acid-and Pepsin-Soluble Collagens from Spines and Skulls of Skipjack Tuna (*Katsuwonus pelamis*). *Chin. J. Nat. Med.* **2014**, *12*, 712–720.

Zaïr, Y.; Duclos, E.; Housez, B.; Vergara, C.; Cazaubiel, M.; Soisson, F. Evaluation of the Satiating Properties of a Fish Protein Hydrolysate among Overweight Women: A Pilot Study. *Nutr. Food Sci.* **2014**, *44*, 389–399.

Zakaria, Z. A.; Kumar, G. H.; Jais, A. M. M.; Sulaiman, M. R.; Somchit, M. N. Antinociceptive, Antiinflammatory and Antipyretic Properties of *Channa striatus* Fillet Aqueous and Lipid-Based Extracts in Rats. Methods Find. *Exp. Clin. Pharmacol.* **2008**, *30*, 355–362.

Zasloff, M.; A. Adams, A. P.; Beckerman, B.; Campbell, A.; Han, Z.; Luijten, E.; Meza, I.; Julander, J.; Mishra, A.; Qu, W.; Taylor, J. M. Weaver, S. C.; Wong, G. C. L. Squalamine as a Broad-Spectrum Systemic Antiviral Agent with Therapeutic Potential. *Proc. Natl. Acad. Sci. U.S.A.* **2011**, *108*, 15978–15983.

Zavareze, E. R.; Telles, A. C.; El Halal, S. L. M.; da Rocha, M.; Colussi, R.; de Assis, L. M.; de Castro, L. A. S.; Dias, A. R. G.; Prentice-Hernández, C. Production and Characterization of Encapsulated Antioxidative Protein Hydrolysates from Whitemouth Croaker (*Micropogonias furnieri*) Muscle and Byproduct. *LWT—Food Sci. Technol.* **2014**, *59*, 841–848.

Zhang, Y.; Liu, W.; Li, G.; Shi, B.; Miao, Y.; Wu, X. Isolation and Partial Characterization of Pepsin-Soluble Collagen from the Skin of Grass Carp (*Ctenopharyngodon idella*). *J. Food Chem.* **2007**, *103*, 906–912.

Zhang, J.; Duan, R.; Tian, Y.; Konno, K. Characterisation of Acid-Soluble Collagen from Skin of Silver Carp (*Hypophthalmichthys molitrix*). *Food Chem.* **2009**, *116*, 318–322.

Zhang, B.; Chen, Y.; Wei, X.; Li, M.; Wang, M. Optimization of Conditions for Collagen Extraction from the Swim Bladders of Grass Carp (*Ctenopharyngodon idella*) by Response Surface Methodology. *Int. J. Food Eng.* **2010**. https://doi.org/10.2202/1556-3758.1772.

Zhang, Z.; Wang, J.; Ding, Y.; Dai, X.; Li, Y. Oral Administration of Marine Collagen Peptides from Chum Salmon Skin Enhances Cutaneous Wound Healing and Angiogenesis in Rats. *J. Sci. Food Agric.* **2011**, *91*, 2173–2179.

Zhang, M.; Li, M. F.; Sun, L. NKLP27: A Teleost NK-Lysin Peptide That Modulates Immune Response, Induces Degradation of Bacterial DNA, and Inhibits Bacterial and Viral Infection. *PLoS One* **2014**, *9*, e106543.

Zhang, R.; Chen, J.; Jiang, V.; Yin, L.; Zhang, X. Antioxidant and Hypoglycaemic Effects of Tilapia Skin Collagen Peptide in Mice. *Int. J. Food Sci. Technol.* **2016**, *51*, 2157–2163.

Zhang, Z.; Hu, X.; Lin, L.; Ding, G.; Yu, F. Immunomodulatory Activity of Low Molecular-Weight Peptides from *Nibea japonica* in RAW264.7 Cells via NF-κB Pathway. *Mar. Drugs* **2019**, *17*, 404.

Zhou, P.; Regenstein, J. M. Comparison of Water Gel Desserts from Fish Skin and Pork Gelatins Using Instrumental Measurements. *J. Food Sci.* **2007**, *72*, C196–C201.

Ziegman, R.; Alewood, P. Bioactive Components in Fish Venoms. *Toxins (Basel)* **2015**, *7*, 1497–1531.

Zodape, G. V. Studies on the Antibacterial Activity of Bioactive Compounds of Fish *Tetraodon fluviatilis* of West Coast of Mumbai. *Biomed. Pharmacol. J.* **2018**, *11*, 513–518.

WEB REFERENCES

Anguilla japonica. https://www.uniprot.org/uniprot/J7GIU8.
Anon. https://sharkstewards.org/shark-finning/shark-finning-fin-facts/.
Anon. https://www.fda.gov/food/dietary-supplements.
Anon. http://www.fao.org/3/x3690e/x3690e1d.htm.
Anon. http://www.fao.org/3/x3690e/x3690e1g.htm.
Anon. Fish Entrails and Processing Waste as a Raw Material. https://www.eurofishmagazine.com/sections/fisheries/item/445-fish-entrails-and-processing-waste-as-a-raw-material.
Anon. Kaposi Sarcoma. http://www.ucdenver.edu/academics/colleges/pharmacy/current-students/OnCampusPharmDStudents/ExperientialProgram/Documents/nutr_monographs/Monograph-shark.pdf.
Barman, D. https://www.researchgate.net/publication/257139523_Fish_Derived_Nutraceuticals_and_food_preservatives.
Centrophorus grunulosus. FAO. http://www.fao.org/fishery/species/2835/en.
Clinical Summary: Evaluation of Nutritional Support with Bonito Peptides in Patients with Hypertension: Summary of Clinical Experience. MET1219 9/05, http://www.metadocs.com/pdf/case_studies/MET1219%20Bonito%20Peptides%20Case%20Reports.pdf.
Cyprinus carpio. https://www.healthbenefitstimes.com/carp-fish/.
Desriac, F.; Jégou, C.; Brillet, B.; Chevalier, P. L.; Fleury, Y. Antimicrobial Peptides from Fish. https://www.academia.edu/13627658/Antimicrobial_Peptides_from_Fish.
Entosphenus tridentatus. YakTriNews.com.
Jayasinghe, C. V. L.; Perera, W. M. K.; Bamunuarachchi, A. Influence of Extraction Methods on Quality of Shark Liver Oils. http://www.fao.org/3/a-bm158e.pdf.
Wikipedia. Health Benefits of Shark Fins.
Kuang, H. K. Appendix III: Non-food Uses of Sharks. Shark Processing Wastes and By-products. http://www.fao.org/3/x3690e/x3690e1d.htm.
Lampetra fluviatilis. *The Druggists Circular and Chemical Gazette*, 1878.
McGrouther, M. Reef Stonefish. https://australianmuseum.net.au/learn/animals/fishes/reef-stonefish-synanceia-verrucosa-bloch-schneider-1801/
Miichthys miiuy. https://www.uniprot.org/uniprot/G0Z5J6.
Mobula hypostoma. Factsheet for the 17th Conference of the Parties (CoP17) to the Convention on International Trade in Endangered Species (CITES) http://www.shark-advocates.org/pdf/facts/cites_devil_ray_tact_sheet.pdf.
Shmerling, R. H. Chondroitin and Melanoma: How Worried Should You Be? https://www.health.harvard.edu/blog/chondroitin-and-melanoma-how-worried-should-you-be-2018050913792.
Unidentified Hagfish. Despite Ick Factor, Slime Eel Has Sex Appeal. http://www.nbcnews.com/id/19336602/ns/health-sexual_health/t/despite-ick-factor-slime-eel-has-sex-appeal/#.XjwXZTFKhPY.

Index

For Product Safety Concerns and Information please contact our EU
representative GPSR@taylorandfrancis.com
Taylor & Francis Verlag GmbH, Kaufingerstraße 24, 80331 München, Germany

www.ingramcontent.com/pod-product-compliance
Lightning Source LLC
Chambersburg PA
CBHW060338220326
41598CB00023B/2748